ABALONE

ABALONE

*The Remarkable History and Uncertain Future
of California's Iconic Shellfish*

Ann Vileisis

Oregon State University Press   Corvallis

Library of Congress Cataloging-in-Publication Data

Names: Vileisis, Ann, author.
Title: Abalone : the remarkable history and uncertain future of California's
    iconic shellfish / Ann Vileisis.
Description: Corvallis : Oregon State University Press, 2020. | Includes
    bibliographical references and index.
Identifiers: LCCN 2019054126 | ISBN 9780870719882 (trade paperback) |
    ISBN 9780870719899
Subjects: LCSH: Abalones—California—History. | Abalone
    populations—California.
Classification: LCC QL430.5.H34 V55 2020 | DDC 594/.3209794—dc23
LC record available at https://lccn.loc.gov/2019054126

♾This paper meets the requirements of ANSI/NISO Z39.48-1992
(Permanence of Paper).

Oregon State University
OSU Press

Oregon State University Press
121 The Valley Library
Corvallis OR 97331-4501
541-737-3166 • fax 541-737-3170
www.osupress.oregonstate.edu

Contents

If you don't know history,
it's as if you were born yesterday.

—Howard Zinn

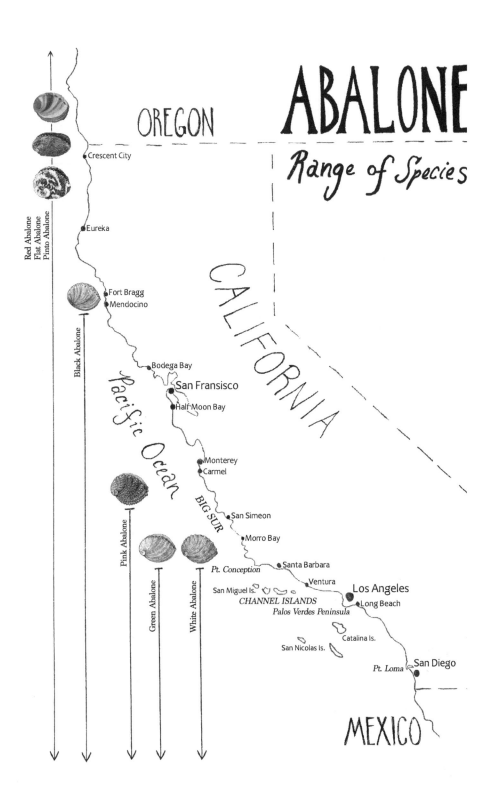

ABALONE
Range of Species

OREGON

CALIFORNIA

MEXICO

Pacific Ocean

BIG SUR

CHANNEL ISLANDS

Crescent City

Eureka

Fort Bragg
Mendocino

Bodega Bay

San Fransisco

Half Moon Bay

Monterey
Carmel

San Simeon

Morro Bay

Pt. Conception
Santa Barbara
Ventura

San Miguel Is.
Palos Verdes Peninsula

Los Angeles
Long Beach

San Nicolas Is.
Catalina Is.

Pt. Loma
San Diego

Red Abalone
Flat Abalone
Pinto Abalone

Black Abalone

Pink Abalone

Green Abalone

White Abalone

Introduction

A MOLLUSK WITH CHARISMA

Ask anyone who spent time on California's coast during the 1950s, 1960s, and early 1970s about abalone, and his or her eyes will likely brighten and then drift to a clear and potent memory: a beachside barbecue, summer vacation foraging in tide pools, or perhaps even plunging into deeper waters to "pop" the meaty shellfish from its rocky dwelling spots. At the time, California's coast seemed awash in abalone. From San Diego to Crescent City, people gorged on abalone steaks, sandwiches, and burgers. Glimmering shells were sold in beach-town trinket shops, nailed to garden fences as signs of diving prowess, and shown off in living rooms as novel décor or psychedelic ashtrays. *Sunset Magazine* routinely showcased abalone recipes and craft projects. Abalones' remarkable abundance and appeal made them icons of California's easy living, laid-back beach culture, and enduring natural opulence.

But by just a few decades later, many younger Californians had never seen or even heard of the legendary shellfish. In a relatively short period, the juicy steaks, ubiquitous shells, and animals that had long studded nearshore reefs vanished from central and southern California, remaining abundant only in northern California, for a time, owing to a tightly managed fishery.

In the past twenty years, two of California's seven abalone species have joined the US endangered species list and four more have become "species of concern." Just recently, even the red abalone in northern California—long regarded as most hardy—has been devastated by a "perfect storm" of environmental stressors, resulting in the closure of the long-popular fishery. The few abalone species that range north into waters from Oregon to Alaska are at risk, too.

After millions of years of gripping tenaciously to North America's western shoreline, after twelve thousand years of sustaining human foragers with their meat and of stirring human imagination with their iridescent shells, how—in our time—did the fate of these delicious, wondrous, and once-abundant mollusks become so precarious?

The answer to that question, and any hope for these animals' future, lies in understanding the remarkable history of people and abalone on the West Coast.

* * *

Consider first that we humans have regarded abalone primarily in relation to our own desires. People have long been intrigued by the captivating beauty of abalone shells and have long savored this creature's meat. Thousands of divers have been drawn to the bracing though sometimes dangerous sport and livelihood of foraging for this shellfish in its undersea haunts. People have loved the selves they become when wearing, hunting, and eating abalone. More than most animals, abalone have been cherished for their rich cultural values.

But keep in mind, too, abalone are also wild animals with their own requirements for survival.

Based on tiny fossils found in ancient marine strata of southern California's coastal mountains, we know abalone have dwelled on the Pacific coast for at least seventy million years. After the notorious Cretaceous-Tertiary mass extinction event wiped out three-quarters of the Earth's plant and animal species, including the dinosaurs, abalone remained absent from the fossil record for more than fifty million years until the Miocene Epoch, when the distinctive mollusks began to reappear as tiny, uncommon specimens, suggesting that some few pocket populations had managed to persist. About five million years ago, abalone became markedly larger and diversified into more distinct forms, but not until about 2.4 million years ago, the beginning of the Pleistocene, did the species we know today begin to come into existence.[1]

It was during these eons before humans showed up that Pacific coast abalone came into their own. In an ocean full of the hungry ancestors of today's fish, octopuses, crab and, eventually, sea otters, abalone's distinct anatomy evolved. The animal's big muscular foot allows it to withstand the powerful force of waves and swells and also to clamp down hard on rocks and reefs, making its sturdy shell a formidable shield against predators that nonetheless drill and pry with relentless determination. The continual press of predation also consigned abalone to dwell in protective cracks and crevices, where their mostly sedentary lifestyle of waiting for currents to serve up bits and pieces of drifting kelp also evolved. And it may well be that some predators—ancestral sea otters by one theory—also kept populations of kelp-eating competitors, such as sea urchins, in check, thus fostering an abundance of savory kelp in the North Pacific that enabled California abalones to eat well and grow large.[2] It turns out that this mollusk has been shaped not only by what it eats but also by what has tried to eat *it* over the course of evolutionary time.[3]

In the late eighteenth century, the great Swedish taxonomist Carl Linnaeus classified abalone into a genus of its own, *Haliotis,* honoring the long tradition of naturalists, stretching back to Aristotle, who had first named the mollusk for the Greek words meaning "sea ear" (*hali* for sea and *otis* for ear), which perfectly fit specimens typical of Europe, with shells the size and shape of a human ear. Linnaeus didn't yet know about California's salad-bowl-sized red abalones—the largest in the world, with some record-breaking specimens reaching more than 12 inches in length.[4] Later naturalists catalogued abalone as gastropods (*gastro* for stomach and *pod* for foot)—what we more commonly call snails—and eventually identified seven distinct species on the California coast, each living within particular ranges of depth and latitude reflecting fealty to water temperature.

Today we know that abalone survive on the rocky, submerged edges of five continents, and despite their wide range in size, all the world's fifty-six delineated species share a large foot and an oval dome-shaped shell with a distinctive interior glimmer, a subtle flattened spiral, and an elegant line of respiratory pores.

The mollusks breathe by drawing water through these small holes to oxygenate gills inside. They use the same pores to expel water and waste and also to broadcast eggs and sperm, which must meet up in the chancy realm of seawater. Although abalone perceive light through primitive eyes, they know their surroundings through chemosensory means. With a fringe of tentacles that reach out from the edge of its big foot, an abalone can detect the scent of drifting kelp and then reach out to grab and draw the nutritious alga toward its mouth or under its shell to store for a later meal.

Yet all the world's abalone now struggle, because the same traits that enabled them to endure through eons made them vulnerable when too many human foragers tramped into their formerly remote habitats. Abalone's astoundingly strong foot became an appealing chunk of meat; its sturdy shell became a prized material for tools and decorations; and its mostly sedentary lifestyle made the animal defenseless to divers with progressively more efficient and far-reaching technologies and increasingly commercial aims. Other consequential environmental stresses piled on, too, as we'll see.

Over the past fifty years, as abalone populations have markedly declined, it has become more and more difficult to see California's *Haliotis* species outside of an aquarium, even if you are a snorkeler or diver. The endangered black abalone, once prevalent in much of California's intertidal zone south of Mendocino, can now be found only in a few remote spots on the Channel Islands. Green abalone can occasionally be seen in shallow waters at some sites in southern California, while all the others—red, pink, and the less common flat and pinto—dwell in deeper waters, observable by tank or breath-hold divers, who may find them infrequently, camouflaged by a thick cover of barnacles, worms, sponges, and other animals and seaweeds that anchor to the outside

CALIFORNIA ABALONE

Ranging from Southern Oregon to Baja,

most common from Mendocino Co. south

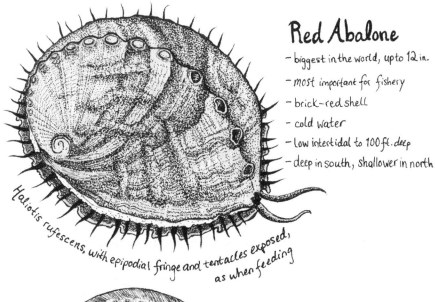

Red Abalone

- biggest in the world, up to 12 in.
- most important for fishery
- brick-red shell
- cold water
- low intertidal to 100 ft. deep
- deep in south, shallower in north

Haliotis rufescens, with epipodial fringe and tentacles exposed, as when feeding

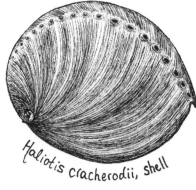

Black Abalone

- up to 8 in.
- smooth black shell
- intertidal
- *ENDANGERED*

Haliotis cracherodii, shell

less common species, not pictured:
Flat Abalone and Pinto Abalone (*SPECIES OF CONCERN*)

Ranging from Central Coast south to Baja
most common from Pt. Conception south

Haliotis sorenseni, shell

White Abalone

- up to 10 in.
- reddish shell, delicate
- subtidal, 20-200 ft. deep
- cold water
- *ENDANGERED*

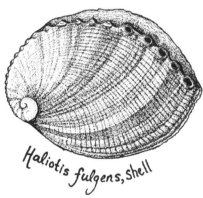

Haliotis fulgens, shell

Green Abalone

- up to 10 in.
- olive green shell, coarse ribs
- subtidal, shallow to 30ft.
- warm water
- *SPECIES OF CONCERN*

Haliotis corrugata, shell

Pink Abalone

- average length: 5.5-6.5 in.
- pink shell with wavy corrugations
- subtidal, 20-80 ft.
- warm water
- *SPECIES OF CONCERN*

of their shells. Whites (also endangered) have scarcely been seen in the wild for decades.

Although biologists place abalone into the category of mollusks—the phylum of life that encompasses invertebrates with soft bodies, ranging from giant octopuses to garden slugs—most people consider them shellfish, a looser cultural category that includes edible mollusks, crustaceans, and some invertebrates. The two words—mollusk and shellfish—reveal a fundamental tension in perspectives that underlies our human relationship with abalone. Mollusk captures how the unique animal fits into the larger system of life on Earth; shellfish refers to the more utilitarian aspects of our relationship—foremost, edibility. The distinction is subtle but permeates this story in profound ways, contributing to cultural confusion that has put the animals' very survival at risk. In short, it's been difficult to think about abalone as both wild animal and meat at once. Yet any understanding of this creature and its plight requires that we do so.

Through time, California abalone have played surprising roles in history at key junctures—as tool, ornament, meat, symbol, and source of wonder and thrill. Abalone have inspired human dancing, singing, and reverence. They abetted colonizers who barely recognized their value, then became a commodity in America's first trans-Pacific trade. Abalone sparked the earliest calls for conservation in California, yet were given reprieve only with the impetus of racial prejudice. Abalone inspired bohemian writers' poetry, California cuisine, and the sport of skin diving, yet also provoked fierce and intractable conflicts over limits on their use. Abalone have driven people to commit crimes (poaching) and to risk their lives pursuing the valuable marine snails even in stormy shark-filled seas.

Meanwhile, given their appeal and increasing vulnerability, abalone have drawn marine biologists on a quest to study the mollusks' cryptic undersea lives in search of answers for how to protect and restore populations in the face of insatiable human demand, devastating disease, shifting ocean conditions, and the ultimate environmental stress—climate change.

Only by knowing a history that integrates these intertwining cultural and ecological threads can we better understand what has happened to the abalones, and why they urgently need our attention now.

* * *

When I began to ask around about abalone, I discovered an upwelling of submerged memories and stories. One neighbor who grew up in southern California in the 1950s fondly remembered foraging in tide pools and then bringing abalone sandwiches instead of peanut butter and jelly for school lunch. A Central Coast friend recalled his chore as a teen was to gather the shellfish for his family's Friday-night abalone dinners. One woman reminisced about heading to the North Coast every summer with her family to camp in the redwoods and dive for abalone that would then be pounded, dredged in cracker crumbs, and fried into cutlets. Several sport divers described foraging for abalone as primal and hallowed. Most abalone stories shared a pattern: vivid physical and sensual experience; deep connection to place, family, and community; and then a wistful dip into nostalgia. Most stories concluded with a poignant feeling of personal loss. Many pointed singular blame at commercial divers, sea otters, too many sport divers, or government mismanagement.

For as long as anyone could remember, and as long as most stories and memories reached back, abalone had been abundant in California, and so it was only natural to consider their abundance as the starting point—the norm—what fisheries scientists call a "baseline." However, the environmental conditions that one generation comes to know and consider as "normal" and "natural" are often, in fact, the result of deeper human and ecological interactions.

This disorienting phenomenon is a central theme in this history of abalone. As we'll see, a cultural amnesia about the historical root cause and ecological context of abalone's past abundance set the stage for a fundamental misunderstanding of the animal's biology that would ultimately lead to its tragic decline. Not until too late did we learn that abalone were more vulnerable to overfishing and environmental fluctuations than we'd realized. Beyond that, the system of state oversight that we relied on to conserve abalone—in the hands of legislators and politically appointed policy makers—for too long proved inadequate to the task.

This human-molluscan history is centered in California, where West Coast abalone were once most prevalent and diverse, and where the culture of their appreciation and exploitation came to its greatest and most-consequential culmination. Within this state there have been three main geographic spheres of abalone life and use, each with its own distinct story: northern California, including the once abalone-rich Mendocino and Sonoma coasts, where local people aggressively fought off commercialization of red abalone to preserve a local culture and economy of sport diving; central California, where the commercial red abalone industry first began, where signs of scarcity first sparked fights for conservation, where abalone became enshrined in California cuisine, and where predatory sea otters revealed the complex realm of community ecology; and southern California, including the Channel Islands, where the greatest numbers and diversity of abalone species once thrived but also where the greatest threats have prevailed, with massive urban populations ever-poised to pollute and exploit the nearshore marine environment. It turns out that stories from each of these three geographies converge in unexpected and momentous ways.

This book is anchored in abalone's cultural significance, but—tracking history—it then kaleidoscopes outward to encompass more biological considerations, as each *Haliotis* species faces its own distinct problems, especially the blacks, whites, and reds. In the chapters ahead, we'll learn more about each of these molluscan characters as knowledge about their biology unfolds. And, as with geography, we'll find that the stories and plights of the distinct species braid together in crucial ways.

In the end, this book recounts how biologists, fishery managers, sport divers, and commercial divers faced up to the fact that, in their lifetimes, abalone went from being plentiful, abundant animals to becoming vulnerable and imperiled. In the crucible of conflict and heartbreak that comes with scarcity, these people found themselves at the very front edge of a profound loss that belongs to us all.

This story of California's abalone ultimately takes us beyond that tragic scenario to the efforts of scientists and citizens who have acted with courage, vision, dedication, and hope to change course and try to save the most imperiled of these species from extinction—and also to the alarming environmental shifts that now bring entirely new challenges to bear.

✳ ✳ ✳

For me, the exploration that led to this book began with the wonder of a single small seashell. I found it while wandering in a small rugged cove on the Big Sur coast. Cliffs towered, cool salt air dampened my face, and giant waves pounded—each one striking a boom that reverberated against the eroding crags. In pauses between the thrusts of seawater, I could hear rocks rumble beneath the outwash of surf. I hopped from one large surf-tumbled stone to the next and then, on a short stretch of beach, a glimmer caught my eye—a lustrous little bowl set in the dark, coarse sand. Less than 2 inches in length, the delicate shell was as thin as porcelain. I turned it over in my hand. The sea had eroded its rough exterior, leaving just a fine form of silvery mother-of-pearl, through and through.

I had no idea how the seemingly fragile shell had landed intact on a beach where pummeling waves jostled bowling-ball-sized boulders like so many marbles. It seemed to be a gift—a precious token to enter an expansive moment of heady wonder. Holding it up to sunlight, I could see the swirls of pale pink, blue, green, and yellow that made up its mysterious iridescence. I marveled. How on earth could an animal create such a brilliant structure—a miniature molluscan version of the Sistine Chapel?

When I found that shell, I didn't know much of anything about the animal that made it. I remembered abalone from a favorite childhood book, *Island of the Blue Dolphins*. I didn't yet know that abalone had for a time grown thick along California's coast. I hadn't yet learned about abalones' ecological relationship with their sea otter predators. I didn't yet know that for decades, perhaps millennia, people had considered hunting and eating abalone an integral part of living on the coast. I didn't realize that two of California's seven abalone species had been listed as endangered. And I wasn't yet aware that abalone may well be the most-argued-over mollusk of all time.

But as a historian of food and nature, I had a strong hunch that the brilliant shell I held in my hand had an important story to tell. That stunning shell inspired me to discover the remarkable history of California's abalone before peoples' cherished memories of them are lost and, hopefully, in time to rouse more interest in urgent efforts toward restoration.

At times, I've wished that I could distill from the abalone shell some telling wisdom, some "gift from the sea," like Anne Morrow Lindbergh famously found beachcombing in the 1960s. Instead, I found a gripping story with swirls of complexity, iridescent facets that can be seen differently in different light. I found both sadness and tragedy in our consequential blind spots and unruly appetites, yet also hope and inspiration in our love of beauty, our connection to place, our hard-minded drive to protect what has become rare, and our persistent search for a path to a future that can be shared with our fellow creatures.

The search for that path is what motivated me to write this book. The brilliant abalone of the West Coast have been too meaningful and too valuable for us to abandon and forget. Their story must show us a better way forward.

PART 1

A Storied Shell

(70 million years BP–1850)

Abalone is the single most important cultural item we have—more than baskets, more than cradleboards. No matter what level a Native person is at in terms of being traditional or not, they always have abalone.

—Tima Lotah Link, Chumash traditionalist, basketmaker,
and educator, interview, March 2014

When you wear it, you are not cheating, not stealing spirit. You are united with spirit: the abalone and the person.

—Florence Silva, Kashaya Pomo elder, quoted
in *Abalone Tales* by Les Field, 2008

With the loss of the abalone, we have lost the ability to feed ourselves with a traditional food that was very nourishing to our bodies. We've lost having abundant shells so that we could make beautiful ceremonial objects or day-to-day objects like beautiful bowls that could be used to drink water. The meat and shells were a big part of our culture for ten thousand years—losing abalone has been a huge loss.

—Mati Waiya, Chumash ceremonial elder and founder of
Wishtoyo Chumash Foundation, interview, April 2011

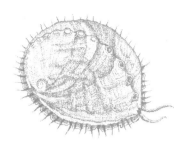

CHAPTER 1
Before the Written Word

For the longest stretch of abalone-human history—the thirteen thousand or more years that indigenous people inhabited California's coast before the arrival of Europeans—we have no written records to tell us what happened; but for parts of the story we can turn to the remarkable record of ancient shells.

Like many animals, mollusks leave behind durable evidence of their existence. Heaps of discarded shells, known as middens, from southern California to the Monterey Peninsula to the San Francisco Bay, attest to millions of shellfish meals eaten by people through the millennia.[1] Mussels, clams, snails, and abalone were crucial foods to the earliest Californians, but the left-behind shells reveal far more than just their role in subsistence.

Nowhere are shell middens so well preserved and chock full of abalone as on California's Channel Islands—remote outposts that sit off the southern coast like an alter ego to the citified mainland. Although most mainland abalone middens have been destroyed by development or disintegration, those on the islands have endured, undisturbed and preserved mostly intact by chalky, calcareous soils.

Abalone shells on the northern Channel Islands provide some of the earliest evidence for human habitation on the West Coast—the remains of the first abalone dinners eaten at the end of the Pleistocene, more than 12,000 years ago. Though Paleoindians likely arrived even earlier, following the Pacific Ocean's kelp-fringed edge from Asia, potential evidence of earlier coastal habitation, especially farther north, was drowned out by a dramatic post-ice-age rise in sea level between eighteen thousand and seven thousand years ago.[2] According to Dr. Jon Erlandson, a University of Oregon archaeologist who has devoted his career to making sense of thousands of years of Channel Island shell midden history, these oldest-known abalone remains turned up high and inland from the sea's edge, near springs and chert outcrops—evidence of small campsites, where Paleoindians made early projectile points. These earliest Californians used abalone as a portable food that could be easily carried, live, in its own shell container.[3]

Think of these isolated ancient shells as the first markers in our human-abalone story. Reading upward through the strata and forward in time, shell middens became larger and more numerous over the course of thousands of years, evidence that growing numbers of hunter-gatherers were staying for longer periods in food-rich locales.[4]

At the same time, the size of the individual shells in the middens shrank, a near-universal pattern that suggests islanders harvested the biggest, easiest to gather shellfish first—and faster than the mollusks could replenish. Intertidal shellfish are especially vulnerable to exploitation because they are visible, accessible, and predictably sit in place, making them easy to take. After the supply of shellfish was exhausted in one place, the islanders moved on to new sites. The periodic shifting of harvest sites allowed for depleted intertidal mussels and black abalone to rebound and become available for another round of foraging. Nevertheless, the trend over thousands of years toward smaller and smaller size shells in middens reveals that even the earliest Californians confronted the challenge of figuring out how to use abalone without depleting them.[5]

Meanwhile, the composition of abalone species in middens changed, too. Reds—the largest and generally deeper-dwelling abalone, owing to preference for cold water—were present early on in late-Pleistocene sites (~13,000 to 10,000 BP, before present), but intertidal black abalone became prevalent in middens dating from the Early Holocene (10,000 to 7,500 BP). Then, in the Middle Holocene (7,500 to 3,500 BP), the red abalone resurged. In the Late Holocene, starting around 3,500 BP, abalone shells of both species became smaller and less abundant, not to be seen again in large quantities until after Europeans arrived thousands of years later, posing one of the most fascinating riddles of Channel Islands archaeology: Why?[6]

Early theories linked changes in abundance of different species either to climate-related fluctuations in seawater temperature or to human taste preferences (in recent times, black abalones were regarded as tougher and less palatable than reds), but neither fully explained the variable shell distribution patterns in the middens.[7]

Then in the 1990s, Jon Erlandson weighed into the unresolved archaeological debate with a new explanation. Drawing on modern ecological science, he and his colleagues recognized that large abalone of any species could exist in appreciable numbers only in the absence of their prime predators: sea otters.[8] The prevalence of large red abalone shells in middens implied that islanders must have hunted sea otters for fur but also perhaps with intent to reduce predation on abalone—a practice that would have allowed the favored shellfish to grow large and abundant in some readily accessible areas, an indigenous shellfish management technique akin to the better-known, ancient "clam gardens" of coastal British Columbia.[9] Evidence for hunting sea otters on the coast dates back at least nine thousand years. At sites on both the Channel Islands and the

1.1. Shell middens on the Channel Islands provide evidence of more than 12,000 years of abalone use by indigenous Californians and reveal critical insights about changes in subsistence, culture, and ecology through time. (Photo by author)

mainland, archaeologists have noted a concomitant increase in both sea otter bones and red abalone shells in middens over time.[10] Erlandson's ecological line of reasoning explained the resurgence of the large red abalone shells, but not why they disappeared from middens during the Late Holocene.

He now thinks beauty may hold the answer. Consider, the millennia spanning the Middle to Late Holocene—when red abalone shell middens waxed to their largest and then waned—was a period when islanders engaged increasingly in specialized crafts and trade. Already practiced in crafting beads from more common *Olivella* snail shells and stringing them into strands for use as currency in California's emerging trade economy, the ancestral Channel Islands Chumash began to make beads and carve decorative pendants from the big, iridescent, red abalone shells.[11] No longer just a by-product of subsistence, shells became a valued material for local use and export. In a world before shiny chrome, silver, and mirrors became commonplace, abalone's iridescence must have been ensorcelling. The largest red abalone middens may be evidence of stockpiling to meet rising demand for brilliant shell ornaments and pendants, which could have driven indigenous ancestors both to hunt sea otters and to wade deeper or dive for red abalone with greater intensity—an interpretation that could explain both the appearance and later dwindling of large shells in the middens.[12] Indeed, evidence not just of whole abalone shells but of artifacts made from them became increasingly common, pointing to the development of a more profound cultural connection between people and abalone that began to emerge during the Middle Holocene.

The earliest known abalone-shell ornaments made by the ancestral Chumash were simple rectangles with two holes in the center, sewn as sequins onto ceremonial garments.[13] Through time, artisans began crafting the shells into pendants for necklaces, earrings, and nose ornaments for personal adornment and using glittering pieces—affixed with asphaltum, a naturally occurring, tar-like glue—to decorate woven-basketry "water bottles," wooden boxes, knives, utensils, and stone mortars used for food preparation.[14] Bone whistles decorated with abalone-shell disks have been found on both island and mainland sites, and abalone-shell inlay would later be used to adorn paddles and seafaring plank canoes, known as *tomols*.[15]

With the growing sophistication of artisans' shell-carving skills and the expansion of trade networks, the world of Native Californians became increasingly spangled with the brilliance of abalone. Shell artifacts were traded first from ancestral Chumash in southern California to ancestral Ohlone and Miwok in San Francisco Bay, and then to tribes in northern California and east into the desert, and beyond. The stunning shell pendants—in an astonishing array of shapes and forms—showed up nearly everywhere indigenous Californians lived, which is to say, nearly everywhere in the state.[16]

THE MEANINGS OF GLIMMER

I met Dr. Ray Corbett in the lobby of the Santa Barbara Museum of Natural History. He ushered me down an outdoor passage, past chirping school children, and into the quiet, dimly lit Hall of Chumash, where orange-backed dioramas present a simplified version of the Santa Barbara area's remarkable prehistory. Corbett, associate curator of archaeology and material culture, agreed to talk with me about the museum's display of abalone-shell artifacts, which span the mind-boggling expanse of more than ten thousand years that indigenous people have inhabited the Santa Barbara area.

The ancestral Chumash used shells first as tools, then as decoration, and, eventually, he told me, "in a role closer to currency"—especially the small *Olivella* snail shells used widely in trade. But abalone was different. It wasn't just "shell money," as some popular nineteenth-century accounts described, projecting American mercantile sensibilities onto Native customs.[17] Abalone likely held special meanings. "With abalone shell," Corbett explained, "artisans spent more time and employed greater craftsmanship, and so we are more likely to see entirely unique artifacts."[18]

The museum's exhibit begins with a showcase of the earliest sequin-style abalone ornaments. Though thousands of years old, the simple pendants still glimmer. Through time, Chumash shell artisans crafted more-ornate punctuated and incised pendants, scratchers, fishhooks, and decorated tools, including knives, water baskets, and bone flutes. Most elaborate of all is a late-period swordfish headdress, set on a mannequin head, adorned with a cascading spray of dozens of abalone spangles that would have shimmered like fish scales when worn in ceremonial dances.[19]

According to Corbett, the abalone shell in its decorative, ornamental function may be best understood as a prehistoric status symbol, not unlike gold or diamonds in our culture. "Gold is valuable because it's shiny, and also because it's rare and not everyone can get it. It was probably the same thing with abalone." He took me over to a diorama illustrating sharp class distinctions that developed in late-period Chumash society. A series of Barbie-doll-sized Chumash couples represent different social classes, with "elites" prominently wearing abalone-shell pendants; the "middle class" dressed with lower-status *Olivella* shell strings; and finally, the poorest "outcasts," with no shell wealth to display. "Not all people had access to abalone in the way that almost everyone—probably even children—could gather mussels. To access abalone, you needed to have physical strength, boats, and tools to pry them off rocks." He showed me a traditional Chumash *tomol*, inlaid with abalone shell at its prow, and then, in a display case, whale rib "pry bars" used to pop abalone off rocks.

Corbett explained that the prestige value stemming from differential access to abalone shells could well have applied to abalone meat, too. "Eating abalone instead of mussels was probably like the difference of eating steak versus hot dogs." Shell midden deposits from the period right around European contact reveal that Island Chumash chiefs held large ceremonial feasts featuring high-prestige foods, including large numbers of abalone. Massive stone mortars decorated with an inlay of abalone shells provide additional evidence for such feasting.[20] Though there is no conclusive evidence for how abalone was cooked, one ethnographic account from San Nicolas Island suggests that ancestral Tongva, the people living south of the Chumash, used large stone pestles and mortars to pound or grind abalone to tenderize its tough meat.[21]

Ray Corbett's interpretation of abalone shell as a rare and esteemed status symbol rests on decades of research and analysis of dozens of California archaeological sites. Over time, abalone seems to have become increasingly valuable. For example, in early Chumash cemetery sites, shell wealth appears to have been equally distributed among families.[22] However, by later periods, abalone pendants tended to be prevalent only in the gravesites of some few male individuals, an observation corroborated by ethnographic and historic accounts of the great wealth and power of Chumash hereditary chiefs.[23] At shell mounds in the San Francisco Bay area, where ornaments were traded from southern California starting in the Middle Holocene, researchers have detected similar patterns and trends.[24] As abalone shells and pendants were increasingly traded, ideas about their value and significance spread throughout indigenous California.

* * *

Much of what we know about the meaning and value of abalone shells to indigenous peoples in ancient times, it should be said, derives from a controversial source. Starting in the late nineteenth century, hundreds of thousands of shell artifacts were collected for universities, museum displays, and art collections by early antiquarians and archaeologists. Self-identifying as explorers and discoverers, they searched for "Indian relics" and were often insensitive to the fact that they were, well, digging up and robbing graves. Though archaeology would eventually develop into a more refined science focused on retrieving information from sites disturbed through construction of road and buildings, Native Americans nevertheless became increasingly incensed by the continuing exhumation of their ancestors.

Starting in the late 1970s, some states, including California, responded to these concerns by passing laws to protect indigenous remains. Mounting indignation finally led Congress to pass the Native American Graves Protection and Repatriation Act (NAGPRA) in 1990, legislation that has since mandated greater attention to the sovereignty and spiritual beliefs of indigenous Americans when gravesites are unearthed. Most tribes prefer to rebury all ancestral remains and belongings as soon as possible and eschew study or display—considering it dishonorable to view abalone artifacts or even photos of artifacts from graves. Tensions remain between the aims of respecting ancestors and of learning more about them, but increasing collaboration between archaeologists and tribal members has led to greater sensitivity and a fuller, more meaningful understanding of the past.[25] Despite their discomfiting provenance, the ornaments and pendants acquired in the century before NAGPRA provide important clues for understanding what abalone shells meant to California's indigenous people before European incursion.

The fact that abalone-shell ornaments have been found predominantly in burial sites underscores their special meaning and role at the end of life. On Santa Rosa Island, ancestral Chumash placed whole abalone shells over the heads of the dead in graves dating back to 7,500 BP, the earliest known ceremonial use of the shells.[26] Similar burials with whole abalone shells, including some atop heads of totemic animals, have also been found at later sites in the San Francisco Bay area.[27] Infants and children were also frequently buried with elaborate garments spangled with abalone-shell sequins, perhaps signifying hereditary status or a ritual associated with the particularly tragic death of a child.[28]

Later ethnographic accounts cast further light on the spiritual significance of buried abalone pendants, suggesting a constellation of related meanings for tribes throughout the state. According to Chumash oral tradition, the soul after death is blinded at a deep abyss and endures great trials before reaching the afterlife, where it is given abalone-shell eyes to see anew.[29] The Yokuts of the San Joaquin Valley held a similar belief and placed disks of abalone shell over the eyes, ears, and mouth of the dead so that they could see, hear, and speak in the afterlife.[30] According to accounts from North Coast tribes, water from an abalone-shell vessel was used ritually to usher the spirits of dead women to the spirit realm.[31]

Abalone played a unique role in transformation not only from life to death but also from childhood to adulthood, especially for women. Throughout southern California, women from different indigenous groups used "scratchers"—elongated pendants made by smoothing and sharpening the abalone's "shoulder." Hundreds of these tools, known in Chumash as *uškik*, were found in late-period burial sites on Santa Rosa Island.[32] In 1884, Chumash descendant Juan Pico explained they were worn around the neck and used by menstruating and pregnant women to relieve itches, following the cultural belief that they'd lose their hair if they touched their heads or skin with their fingernails.[33] Far north, up the coast, whole abalone shells played—and still play—a prominent role in the traditional Hupa ceremony that marks girls' coming of age, serving as a kind of looking glass that reflects a vision for the future.[34]

Beyond associating abalone with passage to the afterlife and to adulthood, many indigenous Californians ascribed the shell with broader supernatural powers as well. Its unique shiny, dazzling, and reflective quality endowed it with special capacities.[35] For example, the reflectivity of abalone shells could be used for healing. Ishi, renowned as the last member of the Sacramento Valley Yahi, used reflected light from an abalone shell to warm rocks to treat an injured friend.[36] The Yokuts wore abalone-shell ornaments hanging from cordage around their necks to reflect sunlight downward to startle rattlesnakes into giving a warning rattle.[37] In the arid mountains of southern California, Kumeyaay shamans used whole abalone shells in rain-making ceremonies, seeking to draw precious water from the skies.[38]

✳ ✳ ✳

Given its beauty, prestige, and perceived powers, it's no surprise that abalone shell was traded beyond California, beginning with small amounts exchanged with ancestral Shoshone people in the Great Basin as early as 3,500 BP. A thousand years later, abalone shells and artifacts began to appear in the desert Southwest.[39]

The liquescent shell became a talisman that shamans and chiefs used for rain-making ceremonies at a time when the region's farmers depended increasingly on erratic rain to water their crops.[40] Archaeologists have unearthed distinctive two-dimensional figurines of shamans and animals carved from abalone shell at thousand-year-old Hohokam sites in southern Arizona.[41]

Although abalone shell has been found at only a few ancient archaeological sites in the interior West, the glimmering ornaments have typically shown up with dramatic prevalence in burials—just as in California.[42] At the renowned Anasazi Pueblo Bonito site—the largest Great House at Chaco Canyon—nearly eight hundred miles from the Pacific, the highest-ranking individuals were interred with numerous abalone ornaments—some perfectly circular in form, some notched into elaborate sunburst patterns, and others irregularly lobed, resembling small puddles.[43]

Abalone shell would later take on sacred meaning for the Navajo and Pueblo peoples as one of four sacred materials associated with their four sacred mountains. The dazzling abalone was associated with the great sunset summit to the west, San Francisco Peak, and appeared in a creation story: the first Navajo woman was made with an ear of yellow corn and an abalone shell.[44] Early twentieth-century ethnographers found abalone shells still used by Zuni rain priests and kachina dancers, and reported that abalone still symbolized "well-being."[45]

Ultimately, fascination for abalone compelled ancient people of the desert Southwest to trade pieces of shell as far east as New Mexico and west Texas, the shell likely becoming more valuable the farther it was carried from its Pacific source. At Ceremonial Cave in west Texas—thought to be a pilgrimage site owing to its cache of ritual items and hundreds of fiber sandals—archaeologists found abalone pendants similar to those made and found in southern California and also a unique eye mask decorated with a mosaic of rectangular abalone tiles.[46]

Centuries later, in 1802, Lewis and Clark would note abalone shell traded as favored "chief's beads" up and down the Columbia River and as far east as Idaho, where the Nez Perce used them to decorate pelt tippets.[47] The Rocky Mountains seem to have created a continental divide for shimmering shells. Beyond that forbidding topographic barrier, indigenous Americans relied on the pearly shells of freshwater mussels from the Mississippi or marine mollusks from the Gulf of Mexico to meet their cultural desires for glimmer.[48]

✳ ✳ ✳

For indigenous southern Californians, the iridescence of abalone shell would prove essential in another way. Starting around 3,500 BP, the shell came to play a new role in subsistence—in the form of glittering fishhooks. Shimmering and strong, the new circular hooks could attract and hold larger fish than the previously used bone toggles designed to catch in a fish's throat.[49] When used together with seafaring *tomols*, the abalone-shell fishhooks enabled growing populations of ancestral Chumash and Tongva to catch previously inaccessible deepwater fishes, increasing the number and type—up to one hundred different species—that could be tapped for food.[50]

The use of these paired technologies intensified during the Medieval Warm Period (800 to 1350 AD), when recurrent and extended droughts challenged islanders. Compelled to stay put in villages near perennial springs, people would have exhausted the nearest shellfish beds—prompting foragers and fishers to look farther afield for different things to eat. When abalone dwindled as a reliable food source, tools made from discarded shells effectively transferred increasing human predation pressure to a new guild of creatures—large fishes, including the abalone-eating sheephead (a type of wrasse).[51] It wouldn't be the only time that human fascination with abalone's glimmer would affect the ecology of California's marine environment in unexpected ways.

Abalone-shell fishhooks would become one of the most common subsistence artifacts found in late-period coastal village sites throughout southern California, their use persisting up to and through the historic period of European contact, when they were noted by several observers, including Spanish soldier Pedro Fages, who in 1769 noted, "The fishhooks are made of pieces of shell fashioned with great skill and art." [52]

Through the Channel Islands' remarkable record of ancient shells, ornaments, and tools, we know that increasing populations of indigenous southern Californians tapped abalone for nutrition, utility, beauty, prestige, and spiritual meaning. Through time, their growing appetites and desire for shell goods surpassed what local abalone stocks could withstand, especially during times of heightened environmental stress. Facing the fundamental human dilemma of scarcity, they found ways to adapt by moving to new shellfish harvest sites, by hunting sea otters, and eventually by inventing new foraging techniques and trading for new foods. However, even with their far-ranging plank boats and bone pry bars, indigenous southern Californians couldn't access all the abalone.

Meanwhile, abalones stuck to their own time-tested strategies for survival. Enough of them hunkered down—hidden and secure in protective cracks and crevices along the coast's rocky shoreline and submerged reefs, in spots with rich, upwelling waters and nearby kelp forests—to parent new generations, continually replenishing, and holding their own.

WHEN SHELLS SING

T'aya, a'ulun, ah' wook, wil, hiwo't, yer'erner', yuxtháran, xōs-saik, Ila'k'waasht'i— these are some of the names that indigenous Californians have used to speak of abalone.[53]

The record of buried shells, ornaments, and tools gives us a coarse sense of how the earliest Californians used and valued abalone through time, but some meanings can be grasped only through words, stories, and ceremonies passed down through generations. Like the scatter of shell and pendants, recorded abalone words and stories are cryptic and few, shared within families or told by only a handful of surviving individuals to some few ethnographers—and only after the brutal disruption of European colonization—yet they reflect backward to deepen our understanding of abalones' profound cultural significance.

Beyond providing meat and shells, abalone appeared in stories that coastal peoples used to explain the creation and order of the physical world.

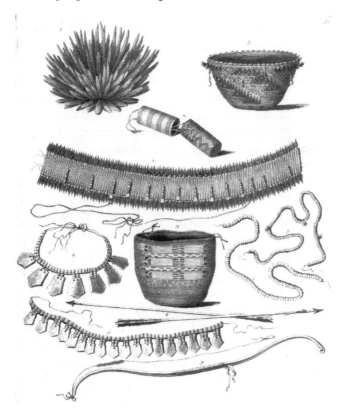

1.2. Drawings by artist George Heinrich von Langersdorff depict two abalone-shell necklaces, plus a basket and other adornments decorated with abalone pendants, showcasing how the iridescent shell was used at the time by Native peoples in northern California. (Plate from von Langersdorff's *Remarks and Observations on a Voyage Around the World from 1803 to 1807*, Bancroft Library, UC Berkeley)

In one Miwok story, the world was dark until Coyote sent men far to the East to capture Sun-woman. Her entire body was covered with *ah'-wook,* the iridescent shells of abalone, and shone so brightly that it was hard to look at her.[54] In a story told by the Wiyot of the lower Eel River, *hiwo't,* (pronounced HAY-WHAT), the abalone, was the "first man" placed into an empty world by the Creator.[55] In the late nineteenth century, when northern California tribes sought to revitalize their disrupted spiritual practices in response to the devastating violence of invading settlers, Kashaya Pomo leader John Boston taught ceremonial dancers to use abalone shell in their regalia because abalone, *wil,* was "the first creature to live in the sea," and, as his granddaughter Florence Silva elaborated "makes all things part of your spirit."[56]

Stories and special songs also informed how and when abalone were harvested for subsistence. According to author and Graton Rancheria tribal leader Greg Sarris, his grandfather Tom Smith, a well-known healer in the Coast Miwok–Southern Pomo community, sang songs before gathering abalone. "You had to know the abalone songs and perform them correctly, otherwise the tide would get you."[57]

One traditional story still told among North Coast's Yurok and Wiyot people emphasizes abalone's enduring association with the female sphere.[58] Fleeing an abusive courtship, Abalone Woman was beaten on the back with a knife. "That's the reason why now, later on, it looks like blood," explained Della Henry Prince, the last fluent Wiyot speaker, describing the dark pink back of a red abalone's shell.[59] In other versions of the story, pieces of iridescent shell still found on the beach are Abalone Woman's tears, and depressions seen in rocks at the ocean—abalone home scars—remain as her footprints.[60] Artist Lyn Risling, whose vivid paintings draw on her Karuk, Yurok, and Hupa heritage, tells the end of Abalone Woman's story this way: "Tears began falling down her cheeks. And as the tears touched the water, the water began to turn all these beautiful colors, and all these shells appeared, and she herself transformed. The young woman transformed into the abalone, and so she became our beautiful abalone shell that we use today for our regalia."[61]

* * *

The word *regalia* carries a particular meaning in describing the shell and bead adornments still used in ceremonial dances of California's North Coast tribes. They are not just beautiful decorations. According to Bradley Marshall, a Hupa artist, scholar, and dancer who makes traditional regalia, once a piece of regalia has been danced, the spirits of the regalia maker and of the animals used enter into it; it becomes its own living being. What's more, the regalia is thought to sing its own songs, a contribution every bit as significant as the shell's brilliance.[62]

Each November, the region's Native peoples—Yurok, Karuk, Hupa, Wiyot, and Tolowa—gather together at the Eureka fairgrounds to honor elders. They welcome the public to watch demonstration dances akin to sacred dances performed as part of their traditional World Renewal ceremonies. When I attended, the crowd was mostly Native, with many people wearing T-shirts that displayed tribal identity under heavy coats left open. Others wore abalone earrings or necklaces; a handful of women had vertical lines tattooed on their chins in traditional manner. We all sat packed on bleachers, chattering and waiting for the dances to begin.

Even before the dancers entered the hall, I heard a showering jangle— thousands of shell pendants sewn to the traditional skirts of the female dancers, knocking against each other with each step, creating a distinctive shush of shell music. As the women and girls followed the male dancers onto the dance floor, the crowd quieted, and the unique clacking shell song crescendoed and filled the large room like an auditory shimmer.

That afternoon, Tolowa dancers demonstrated a Feather Dance. The lead dancer wore a traditional Tolowa headdress, a flat rectangle of buckskin adorned with bright red woodpecker feathers. All the men and boys wore headbands with two large eagle feathers, one over each ear, knee-length buckskin wraps around their waists, and shell regalia draped around their necks—some made from long abalone shoulder pendants strung together with pine nut beads. Several men carried buckskin quivers ornamented with large abalone-shell medallions. The women's skirts were covered with hundreds of shells, including oval-shaped abalone pendants hanging heavy from hems like a glittering fringe.

When the lead dancer began to intone a syncopated rhythm, the line of dancers started to sway back and forth. Traditionally, dancers perform in front of a fire, and the abalone shell's reflection of firelight plays a special role in chasing out bad spirit. Here, under fluorescent glare, the pendants still caught light and glinted with the dancers' sway, and the chorus of clacking shells added its spirit-infused music.

Women's ceremonial skirts and aprons count as regalia, too. After the dances ended, I talked to the mother of a young girl who'd danced for the first time, wearing a small, shell-adorned skirt. The mother told me she had learned traditional dances as a child and wanted her daughter to have the same experience.

I asked her about the dancers' shell-covered skirts. "Once a skirt is made," she explained, "it is considered to be alive and to have a life of its own." Some shell skirts have been kept in families for decades, carefully tended and repaired as need be. With renewed efforts to revitalize traditional dances, she told me, she'd been asked to join a group of Tolowa women who are making new dance skirts to accommodate more young dancers. In addition, I later learned that the Native Women's Collective, a grassroots group that fosters growth of indigenous arts and culture, has been collecting photos and personal stories about women's dance regalia and sharing them in an online exhibit—the Northwest Coast Regalia Stories Project—a good place to learn more about this continuing tradition.[63]

The shimmering, singing ornaments in the traditional Tolowa dance regalia represent a profound symbolic and spiritual realm of indigenous abalone-shell use that stretches back thousands of years and spans cultural groups up and down the coast. It's no surprise that abalone shell remains a sign of cultural continuity, endurance, and renewal.

✳ ✳ ✳

Starting with simple camp meals, when abalone provided meat to California's earliest Paleoindians through thousands of years of cultural evolution and elaboration, the iridescent-shelled abalone offered hundreds of generations of Native Californians nutrition, tools, beauty, kinship, prestige, creativity, strength, succor, ceremony, and a connection to supernatural realms—all the while managing to persist in nearshore coastal waters.

The close relationship between people and abalone would shift in consequential ways with the arrival of European colonizers.

CHAPTER 2
The Unexpected Consequence of Shimmer

Father Eusebio Kino first laid eyes on the brilliant shells when he arrived at the Pacific coast in 1684. He was serving as Jesuit missionary and cartographer with the Atondo Expedition. After efforts by conquistador Hernando Cortez had failed to establish a permanent settlement for Spain in Baja California, Atondo and Kino now attempted the strategic task.

From mainland Mexico, they'd sailed across what is now called the Gulf of California to the Baja Peninsula. Then, for three days, the party had crossed arid, spiny, unforgiving mountains looking for arable land and fresh water that might supply a new mission. They found none, but when they reached the Pacific Ocean at low tide, members of the party discovered "shells of rare and beautiful luster, of the colors of the rainbow, every one of them larger than the largest mother-of-pearl shells."[1]

Fifteen years later, hundreds of miles east from the coast, Kino was working his missionary aims at the northern frontier of New Spain, having arrived in the desert borderlands of present-day Arizona from Sonora, Mexico. Quechua people (Kino called them Yuma) in the valley of the lower Colorado River gave him some "curious and beautiful blue shells." With a flash of insight, he realized these were the same shells he'd seen years earlier on the Pacific coast. Their unforgettable luster had lodged deep in Kino's memory, but it was their promise that now excited him.[2]

To Kino, the presence of shimmering "blue shells" in Quechua country affirmed the likelihood of an overland route to California and the Pacific. Thus far, Spanish explorers had reached California only by sea, and many European cartographers remained unclear about whether California was an island or a peninsula.[3] Some of the confusion stemmed from an earlier account of abalone shells. In 1604, conquistador Juan de Oñate had noted stories from the lower Gila River nations about an island to the west where all the inhabitants wore "around the necks and in the ears pearl shells." Likely they referred to the ancestral Tongva, who inhabited Catalina in the southern Channel Islands chain.[4] Adding to the puzzlement, a popular 1510 Spanish romance novel had extolled a mythic "island of California."[5] However, if California was indeed a peninsula attached to the mainland, Spain could deliver cattle overland from its rancherias in Sonora to supply galleons returning from the Philippines and new missions on the coast. If abalone shells could help Kino find this overland route, Spain could settle all of California, enabling the Padres to bring more indigenous "souls" into their Christian fold.[6]

Kino traveled to a rancheria at an ancient crossroads of Native trails near present-day Tucson and sent out messengers in all directions to inquire about the "blue shells." Within a week, they returned with emissaries from many tribes, some from more than a hundred miles away. Deep into the night the men talked around a fire, Kino wrote, about "the eternal salvation of all those nations" but also extensively about "the blue shells which were brought from the northwest and from the Yumas and Cutganes."[7] Of course, at that point, indigenous shell trade from the Channel Islands through the Southwest had persisted for more than one thousand years. But to Kino, it seemed a new discovery when the Native peoples confirmed that the shells came only from a larger sea on what Kino called "the opposite coast of California," a ten- or twelve-day journey beyond the Gulf of California.[8]

After Kino first took notice of the blue shells, he started to find them in settlements everywhere he traveled, as if his eyes homed in on their watery appearance in the arid landscape. People from many Colorado and Gila River tribes gave Kino blue shells as gifts. Again and again, the shells appeared in Kino's diary and letters, as he built a mass of evidence for the land passage. He hoped the iridescent shells could lead him, like so many marks on a treasure map, to discover its path.[9]

In the fall of 1700, Kino finally set out in search of the shell-inspired land passage.[10] Traveling down the Gila River, he returned to its confluence with the Colorado, where the Quechua (Yuma) had first given him abalone shells.[11] When Kino climbed a hill to survey with his telescope, he could see a broad flat expanse of land and, he thought, California in the distance—with no sea "between and separating." He was overjoyed but could go no farther. His Native guides were "weary" of travel, and it would soon be time to collect "alms of cattle" back in Sonora, he wrote, to explain his reason for turning back.[12]

In the spring of 1702, a companion joined him on his quest. "So great was our desire . . . to cross to the Sea of the West," the yearning Kino wrote, but this time, his friend fell ill and the horses had trouble crossing the enormous wetlands of the Colorado River estuary.[13] Again, he was compelled to turn back. All in all, he attempted fourteen expeditions to find the overland route, but he never made it back to the Pacific—the source of the abalone shells that had captured his imagination.[14]

Nevertheless, Kino's obsessive promotion of the potential route would ultimately prove consequential. Seven decades later, this pathway of the ancient shell trade became the route Captain Juan Bautista De Anza followed when he led soldiers, settlers, and cattle overland for the first time to bolster Spain's new missions in California.[15]

With the founding of the missions in the 1770s, Spanish colonization would begin to devastate Native Californian communities with deadly diseases, brutal treatment, and ravaging livestock that lay waste to the landscape, destroying traditional plant foods, usurping game habitats, and undermining long-standing indigenous subsistence practices.[16] By the mid-nineteenth century, the estimated precontact population of more than 310,000 Native Californians would plummet to less than half.[17]

Through this period of tragic upheaval, abalone would continue to play customary roles, even as Native people were forced to adopt mission ways. In many places, indigenous people continued to forage for the shellfish, and Ohlone and Yokuts at Mission Santa Cruz continued use abalone-shell fishhooks. At missions in San Francisco and San Jose, the Ohlone persisted in their ceremonial dances, as did the Luiseño at Mission San Luis Rey—continuing to wear traditional regalia made with abalone shells. A remarkable mother-of-pearl "tabernacle" made from painted wood, mirrors, and abalone shell in the 1790s at Mission Santa Barbara affords rare evidence that Chumash mastery with shellwork persisted in the mission context. Finally, the Ohlone at Mission Santa Clara continued to practice traditional burial customs even as missionaries pushed new spiritual beliefs.[18]

Surviving indigenous Californians would continue to use and cherish abalone, but with the increasing dominance of the Spanish in California, the center of our story shifts as the new colonizers soon found their own use for abalone.

✳ ✳ ✳

Though Spanish mariners and explorers had made occasional stops on California's coast long before settlement, they'd made only passing note of abalone. In 1587, when Captain Pedro de Unamuno landed his galleon near present-day Morro Bay, he planted a cross in a large pile of abalone shells, instigating an ominous and deadly skirmish with local Chumash residents. In 1602, when Father Antonio de la Ascensión arrived in Monterey with the Vizcaino Expedition, he observed the "large shells fastened to the lowest parts of the rock" and noted how Native Rumsen Ohlone people used their meat and "bright" shells.[19]

Not until the late eighteenth century did the Spanish recognize that the shimmer of abalone could become a path to their own wealth. On a beach near Monterey, the glint of shells caught the eyes of visiting sailors in 1774. They carried the iridescent shells aboard their vessel *Santiago*, which was headed north under Captain Juan Jose Perez to explore and patrol the Pacific Northwest coast. Though Spain had enjoyed some two hundred of years of exclusive maritime dominance in the Pacific, Perez was now scouting for incursions by British and Russian mariners who were increasingly entering its waters.

As they sailed north, the crew of the *Santiago* bartered with Native peoples. On Vancouver Island, when sailors arrayed their beautiful Monterey abalone shells, the local Nuu-chah-nulth people (known historically as Nootka) were smitten and offered their finest furs—thick sea otter pelts—in exchange. The local northern or pinto abalone, *Haliotis kamtschatkana*, was present in this area, but they are smaller and less vividly iridescent than California's red abalone. In his account of the journey, Father Juan Crespi noted, "These Indians had a great liking for these shells."[20] Thus the West Coast fur trade began with the compelling glimmer of abalone.[21]

On the next two Spanish voyages to the North Pacific, crewmembers tried to barter for sea otter pelts with more typical trade goods—beads, knives, old clothes, and pieces of iron—but Monterey abalone shells remained by far the most coveted items.[22] The Vancouver islanders used the brilliant shells as lavish containers, earrings, and ornaments on special blankets and hats, while shaman artisans worked pieces of shell into haunting dance masks, the iridescent inlay giving eyes of the totemic animals an otherworldly glow.[23] Soon every Spanish ship headed north from California loaded up with abalone shells in Monterey. On his 1789 voyage, sailor Esteban José Martínez wrote that the Nootkans "valued the large shells as much as if they were gold or silver."[24]

The Nuu-chah-nulth's excitement about California abalone shells fit into broader global designs and desires for beauty and prestige. In China, the rising class of mandarins had learned of the beauty and value of lustrous Siberian otter pelts through trade with Russia and paid premium prices. In Spain, the royal court continued to covet silver and gold, but New Spain's mines were handicapped by insufficient stores of quicksilver, or mercury, needed to process the precious metal ores. When Spanish sailors found that they could trade seashells—readily picked up "for a trifling" in Monterey—for valuable otter skins, they discovered the key to a new Pacific trading scheme.[25]

They ferried otter pelts back to the port of Acapulco, where Spain's fat-bellied galleons set sail on equatorial winds to what is now the Philippines. There, the furs were traded for Chinese quicksilver, which was brought back in jars and then hauled overland to Spain's New World metal mines. In the meantime, smaller ships ran back up the coast to pick up more abalone shells to trade for more furs to feed the trans-Pacific trade.[26] Spanish mercantilists made great fortunes turning the glimmer of seashells into the glimmer of quicksilver and, ultimately, the glimmer of gold.

The trade in shimmer had devastating effects. Spanish sailors carried tuberculosis, syphilis, and smallpox, unleashing calamitous diseases that decimated Native villages of the Northwest and eventually the entire West Coast. Offshore, too—within just a few decades, sea otters of the Northwest coast were nearly wiped out, prompting Russian traders with enslaved Aleut hunters to move south into California's rich waters, where they joined Spanish, British, and, eventually, American ships in hunting sea otters along the mainland coast, around the Channel Islands, and in Baja.[27] Historian Adele Ogden, who authored the classic account of California's sea otter trade, tallied up records from more than a hundred European and American ships engaged in the fur trade through 1830 and determined that about fifty thousand otter pelts were traded in just five decades.[28] The hunt would continue, but the haul of pelts would inevitably decline from thousands to hundreds, and then to mere dozens, until California's sea otters—with an estimated historic population of nearly sixteen thousand animals—were almost entirely exterminated, too.[29]

The massive sea otter hunt, sparked by the cultural appeal of abalone shells, would unwittingly transform the underwater world of the very mollusk that created the alluring shells. With their top predator plucked out of the Pacific's nearshore waters, the otter's prey—invertebrate animals including abalone and sea urchins—would flourish in an entirely unprecedented manner. Abalone grew larger and spilled out into habitats beyond the cracks and crevices that had long sheltered them, with some species piling one on top of another. Within a few decades, they would become superabundant.

However, this immense intertidal and undersea shift would take more than a century to decipher. Imagine great horizontal arrows of the Pacific shimmer trade intersecting with arrows of undersea food webs that Californians wouldn't even begin to grasp until the 1960s, when the technology of scuba diving and the science of ecology would open up the possibility for new underwater research and understanding. The only people who would have noticed and passed along stories of the massive ecosystem change—the indigenous coastal tribes—were wracked by disease and violence long before the descendants of their conquerors would recognize the importance of learning about the culture and nature of the place.

I did manage to find one secondhand indigenous account of this pivotal ecological transformation. It was recorded in the 1850s by Monterey news reporter Alexander Taylor, one of the first Americans to write about surviving indigenous cultures with interest. Taylor was the first to publish the Native Rumsen Ohlone word *a'ulun* (pronounced ow-loon) as "aulone," which slipped into widespread usage around Monterey, ultimately transmogrifying into the American English word "abalone."[30]

In an 1856 article in the *Sacramento Daily Union*, Taylor recounted a story told to him by an "Old Monterey Indian." The man's father used to say that "the Aulone formed a kind of epocha [*sic*] for his tribe, as it was only found by the Indians around Monterey Bay, not many years previous to the arrival of the Spaniards."[31] Though it's difficult to know precisely what he was referring to, the father clearly noted an epoch or time marked by abalone.

This fleeting story is a clue that the prevalence of abalone on Monterey's rocky headlands was not as natural or timeless as it seemed to the newcomers. Yet to increasing numbers of European and American mariners who arrived in California in the nineteenth century, abundant abalone seemed to be a "celebrated" and foundational part of the natural order of the coast.[32] Their first encounters with the brilliant shellfish would become a new starting place—a new cultural baseline for thinking about abalone.

CHAPTER 3
The Anatomy of Iridescence

The red abalone shell on my desk looks like a hammered silver bowl, but if I move it slightly, its colors change seductively. If I look down on the shell, it's pink, but if I hold it up to my eyes, the pink swaths turn gold and then deep blue-green. Even when I hold the shell still and steady, its surface dances with colored light. The pastel luster glows, popping out with a luminous life of its own.

This is the intriguing shimmer that has drawn people to pick up abalone shells and carry them halfway across the continent, indeed, halfway around the world. Commonly called "mother-of-pearl," the distinctive glimmering material on the inside of abalone shells is more precisely called "nacre" (pronounced NAY-ker)—an abrupt-sounding word for something so lustrous and smooth.

To understand the wondrous glimmer of an abalone's shell, you must think layers, and then layers within layers. In just the past thirty years, scanning electron microscopy has revealed that nacre is composed of layers of tiny polygonal tiles of aragonite, a crystalline form of calcium carbonate. Each tile is roughly 10 by 7 micrometers and only 0.5 micrometers thick.[1] To get a sense of the minute scale of these tiles, imagine that more than 150 could fit on the tip of a human hair! The tiles are neatly stacked into tessellated columns, like tiny frosted-glass bricks held together with a mortar of thin organic glue. Ironically, it is this highly organized crystalline microstructure that gives abalone shell its freewheeling colors.

As I watch the play of light on shell, I can't help but think that this iridescent vessel holds the essence of the ocean's beauty—as if the animal that made the shell distilled silvery splinters from its watery home and then crystallized them, like a painter catching on canvas fleeting diamonds of light that glitter on the sea. As it turns out, this is not far from the truth.

* * *

The mystery of seashells' beautiful iridescence first came under scientific scrutiny long before the invention of electron microscopes. At the outset of the nineteenth century, the Scottish physicist David Brewster became fascinated by the play of light on mother-of-pearl. The way the colors glinted and changed when he moved a pearl-oyster shell seemed a form of "natural magic." Through a noted career, Brewster would go on to discover the "Brewster's angle," known to students of physics; to invent the kaleidoscope; to write a definitive biography of his hero, Sir Isaac Newton; to author his own *Treatise on Optics*; and to be knighted.[2] But early on, intrigued by the "splendid exhibition of colours . . . and the successive development of fresh tints, by every gentle inclination," the twenty-something Brewster stumbled somewhat accidentally onto the hitherto "unknown and extraordinary cause" of seashells' iridescence.[3]

Conducting optical experiments with the light of a candle, he analyzed how the colors changed when he observed light reflecting on the shell surface from different incoming or incident angles. To measure the angles exactly, Brewster had glued his mother-of-pearl to a goniometer—a precision rotational instrument—with a "cement" of rosin and beeswax. When he finished his measurements and removed the shell from the device, he was surprised by an unexpected discovery. The shell had left "a clean impression of its own surface" on the cement, he wrote, which itself had "by this means received the property of producing the colours which were exhibited by the mother of pearl."[4]

Incredulous, Brewster tried again, and again, and found he could impart iridescent properties to all manner of impressionable substances—from red wax and balsam of Tolu to gum arabic, mercury, bismuth, and more. As he explained it, "The mother of pearl really communicated to the cement the properties which it possessed."[5]

Examining the shell with a microscope, Brewster was doubly surprised to find that he could discern an "elementary grooved" pattern in its apparently smooth surface.[6] In some areas of the shell surface, he painstakingly counted three thousand grooves to the inch. Then, he found the same minute corrugations in the wax cement impression.[7]

With these unexpected revelations, Brewster recognized that the capricious colors he observed with mother-of-pearl resulted from the diffraction of light bouncing off the microscopic "eminences and depressions" in the shell's surface.[8] These are what we now call "structural" colors, as distinguished from the more common, pigment-derived colors of fruits, cloth, and flowers, in which wavelengths of light are differentially absorbed and reflected. (Grass is green, for example, because chlorophyll pigments within it absorb all the colors of the light spectrum except green, which is reflected back to our eyes.)[9]

Modern-day optical scientists call the minute grooves that Brewster identified "diffraction gratings." Light hits the parallel surfaces of the fine gratings, and if the size of the wavelengths is just right relative to the spacing of the miniature canyons, the light will be split into its component colors and diffracted in different directions. The undulating light waves then "interfere" with each other—either "constructively," by coalescing into the same phase, which intensifies the color, or "destructively," by knocking each other out of sync, effectively shrinking the waves and diminishing the color.[10]

Some of the colors Brewster saw in his mother-of-pearl sample—especially the brilliant crimson visible at low angles—were not transferred to the wax mold. These "non-transferable colors," he ascribed to additional diffractions and reflections occurring within the shell's thin pearly layers, an optical phenomenon that scientists now call "layer diffraction." By Brewster's account, both the surface grating diffractions and the internal layer diffractions accounted for the iridescent colors that seemed to glow from within the mother-of-pearl.

Remarkably, with an early microscope and candlelight, Brewster established the basis for a theory that continues to be refined by modern-day optical scientists now working with electron microscopes, laser beams, and more distinctively nacreous abalone shells.[11]

Mineralogists in Australia have observed the microstructure of abalone nacre and found minute aragonite tiles arranged one atop another in tidy columns. They regard the uniform height of the tiles to be the key factor in producing abalones' uniquely vivid iridescence. As light penetrates the neatly stacked, translucent tiles to depths of many layers, it is diffracted, creating a kaleidoscope of interference. Because the thickness of the tile layers is similar to wavelengths of visible light, the waves fall readily "into phase"—the sweet spot of color reinforcement that gives rise to brilliance.[12] The mineralogists also found that shells of different abalone species had slightly different tile heights, explaining the physical reason that shells of some have a pink cast while others, like New Zealand's pāua, look purple.[13]

It's because the aragonite tile layers are so thin that the appearance of abalone nacre can change with even just a slight movement. When I lift the shell from my desk to eye level, all the incident angles of light shift, and so a different set of light waves might be in sync and reinforcing color, or out-of-phase and cancelling color—in either case, completely changing the appearance of its iridescence. That's why looking at an abalone shell is so captivating.

* * *

On first glance, it would seem that the visual extravaganza of shell iridescence must serve to attract some beholder of beauty. Evolutionary biologists have found that the colorful shimmer of peacock and hummingbird feathers—created by gratings and tile layers similar in structure to those found in abalone shells—confers advantage in mating and therefore has intensified through sexual selection over time.[14]

However, there is no corollary explanation for the beauty of abalone shell, as there is no element of visual attraction in the broadcast spawning that is abalone sex, and the nacre of abalone remains hidden inside until after the animal dies. The iridescence of nacre has absolutely no visual value for the mollusks that make it.

Most likely, it's a dazzling side effect—what paleontologist Stephen Jay Gould and geneticist Richard Lewontin famously called a "spandrel," using the architectural feature from Renaissance cathedrals as a metaphor to describe traits that appear to have one purpose, such as adornment, but, in fact, serve other, often-structural functions.[15]

Indeed, the layered tiles that make for abalone shells' brilliant shimmer also make for their strength. Abalones need strong shells to protect their large, meaty bodies. The shells' unique microstructure evolved in an ocean full of voracious predators.[16]

As it turns out, abalone shell is one of the strongest natural materials on the planet.

* * *

Dr. Marc Meyers opened his top desk drawer and took out a piece of chalk. "Abalone shell is made up of calcium carbonate," he explained. "In its pure form, carbonate—like this chalk—is extremely brittle." He broke the chalk for emphasis. Then he reached for an abalone shell propped on his windowsill and wrenched it in his hands, but to no effect. "Because the abalone shell has a very small amount of organic material—just 5 percent—it is a thousand times stronger than pure carbonate. I can't break it no matter how hard I try. Basically, the abalone has optimized the mix of organic material and carbonate in its shell over the course of millions of years to create the strongest structure possible."

Meyers is a materials scientist in the University of California San Diego's Department of Mechanical and Aerospace Engineering. As a schoolboy, growing up in Brazil, he snuck into a nearby steel mill with other kids delivering lunch pails to their fathers. There amid deafening machinery, Meyers saw steel flowing hot and red, bursting in sparks, but then turning solid, gray, and hard. Around the same time, while hunting in the nearby rain forest, he found the bill of a toucan. When he picked it up, it weighed next to nothing, yet it, too, was remarkably tough. Though the properties of steel and other metals guided Meyers's career for decades, the weightlessness of the toucan's bill stuck in his memory. After turning fifty, he applied his scientific savvy to investigating natural materials. Meyers has since studied toucan bills, piranha teeth, armadillo armor, elk antlers, crab exoskeletons, and, ultimately, abalone shell—all with an eye toward industrial applications. In his lab office, Meyers explained to me his fascination for biomimicry: "Nature has already tried and tested designs over the course of millions of years. The lessons are there. It's up to us to learn them."[17]

Meyers has subjected abalone shells to compression, pulling, and even gunshots, and then examined the patterns of fracture at a microscopic level to figure out why the nacre is so tough.

The answer lies in the shell's microstructure, which can be seen only with a scanning electron microscope (SEM). Meyers shuffled through papers on his desk to find some striking black-and-white SEM photos that showed neatly stacked aragonite tiles in cross section from a fractured edge. He pointed out that the tiles themselves didn't break; the mortar broke, and the tiles pulled apart. One of the key things that Meyers and other materials scientists have determined is that the "brick and mortar" microstructure of the shell helps deflect cracking by dissipating energy to the organic layers, which serve as a viscous glue that not only holds the tiles together but also has some give to let them slide slightly apart. This gives the animal a chance to repair its shell.

Though Meyers started out looking for natural structural models he could mimic to develop tougher military armor and ceramics, he became engaged in a broader project that encompasses the work of dozens of biochemists and molecular biologists. Starting in the 1990s, it became somewhat of a holy grail in the budding field of nanotechnology to understand how abalones build their highly ordered nacre on the molecular level. Typically, industrial manufacture of composite materials requires tremendous force, energy, and high temperatures. But abalone readily assemble their durable shells at ambient temperature and pressure, simply using organic macromolecules to catalyze and crystallize minerals from the basic elements of seawater. If the concept of biomimicry could be extended to duplicate a bottoms-up approach to manufacturing, the applications would be revolutionary.

Efforts to decode the abalone's shell-making secrets have made abalone nacre one of the most studied natural composite materials in the world. Researchers using proteins extracted from nacreous shell material have been able to catalyze the construction of carbonate in vitro, with biomedical applications for artificial bone and teeth already in the pipeline.[18]

Back at my desk, I download some of the SEM images that Meyers has published and stare at them on my computer screen. Looking at the black-and-white photos of the perfectly stacked tiles, I am transported to an entirely different universe of the abalone shell. In my mind's eye, I try to envision at

3.1. Scanning electron microscopy reveals the remarkably ordered tile microstructure that makes the strong and beautifully iridescent nacre inside abalone shells. (Photo by Marc Meyers)

once all the tiny tiles, the potent protein glue, the light waves ricocheting off the many internal layers to create the chaotic dance of color. All this occurs at an unfathomably infinitesimal scale and yet creates such tangible strength and beauty. I reach for the sturdy red abalone shell on my desk and hold it in my hands, rocking it slowly back and forth to kindle its mysterious glimmer with new appreciation.

PART 2
The Abalone Century
(1850–1962)

There is a family of Mollusca whose beautiful shells are frequently seen ornamenting the parlor mantel or centre table, the admiration of all on account of the brilliant colors and iridescence of their pearly interiors.

> —Robert E. C. Stearns, 1869, *American Naturalist,* July 1869

From a want of the requisite knowledge to prepare them for use, aulones are seldom found on our tables; and when occasionally an enterprising lady by way of variety, sets a dish of them before her friends, satisfaction is rarely given the eater on account of their insipidity and toughness.

> —*Monterey Weekly Herald,* October 3, 1874

Get the aulones fresh in the shell and alive. With a common hammer beat on the face of the fish . . . until they quit their hold and drop from the shell. . . . Carefully wash away everything but the pure white fish; then wrap them up in a piece of strong cloth, and with a billet of wood, pound them to a jelly. Boil the jelly for a quarter of an hour, with milk and butter, peppered to suit the taste, and a dish to satisfy the most fastidious, will be the result.

> —the oldest printed recipe for California abalone I could find, *Monterey Weekly Herald,* October 3, 1874

When I was a boy in Pacific Grove in the late 1880s and 1890s, the abalone were so thick on the shoreline rocks that you couldn't walk out at low tide without stepping all over them. In some places . . . abalone would sometimes be several deep, covering whole crevices of rock canyons and extending over 100 feet out into the water when exposed at low tide. . . . Anybody, without any experience, could gather all the abalones he wanted without getting his feet wet.

> —W. R. Holman, age ninety-five, remembering his childhood with John Woolfenden, *Monterey Herald,* June 13, 1976

For three or four years past the business in these shells has been very extensive; but fears are felt for its future, since the mollusks are being rapidly exterminated along the whole coast.

—Ernest Ingersoll, US Fish Commission, quoted in Goode, *The Fisheries and Fishery Industry of the United States*, 1887

Following along the rocky shore one observes this sign: "Abalone Trail," but as a rule, look as you will, you will find no abalones, so complete has been the destruction.

—John Oliver, 1916, *California Fish and Game*, 1916

Experience has shown that many of the problems of fish and game management are not best decided by a legislature. What is needed is a stable commission, free from political pressure and upheavals, endowed with sufficient regulatory powers to adopt and carry out those conservation measures which are based on technical investigations.

—N. B. Scofield, *Thirty-Fifth Biennial Report of the California Fish and Game Commission*, 1936–37

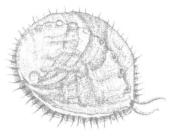

CHAPTER 4
Becoming a Commodity and an Icon

While the hype and fever of California's 1849 Gold Rush drew most treasure seekers toward the mountains, a more obscure rush began on the coast with just as much excitement. Thousands of immigrants, fleeing violent rebellion in Guangdong Province on China's southeast coast, had crossed the Pacific to California, lured, like so many others, by the promise and glimmer of gold. But it was on wave-blasted rocks that a small group of Chinese fishermen found their unexpected treasure: seemingly infinite quantities of abalone.[1]

In China, abalone had long been regarded as a luxury reserved for the wealthy.[2] Since the Zhou Dynasty (1046–256 BCE), artisans had used shimmering nacre as inlay in ceremonial lacquer pieces.[3] Fascination for the shellfish grew in the second century BCE with Qin Shi Huang, China's first emperor, best known now for the massive terra-cotta army he had built to safeguard his crypt. Obsessed with attaining immortality, Qin quested after fabled elixirs, including abalone from Japan's rugged east coast. Later, the shellfish became an important form of diplomatic tribute, with Korea and Japan routinely sending abalone to show deference to China's rulers.[4] Abalone showed up on the menus of grand imperial feasts as one of the "four treasures." Because the Chinese word for abalone, *baoyu* (pronounced BOW-you), sounds like a phrase that means "guaranteed wealth," abalone meat came to symbolize good tidings and abundance.[5] More important, the Chinese have long considered abalone to be a tonic food—its rich meat serving to restore lost qi, or energy—and many have believed its meat has aphrodisiac potency.[6] Even today, abalone remains a highly coveted ingredient for Chinese wedding soups and New Year's feasts and, in dried form, can sell in San Francisco's Chinatown for $2,000 per pound.[7] Moreover, in traditional Chinese medicine, powder ground from abalone shell has long been considered as a remedy to improve vision.[8]

To meet the demand created by such high esteem, centuries of heavy fishing had emptied China's rocky shorelines of abalone. By the mid-nineteenth century, Chinese law prohibited peasants from gathering *baoyu*.[9] Given this context, it must have been a sweet eureka moment when those mid-nineteenth-century Chinese immigrants discovered abalone growing large and thick along California's rocky coast. And what's more, Americans seemed mostly to ignore the prized shellfish.

These were, of course, the abalone released from predation to flourish in the wake of the fur trade several decades earlier. The untapped economic and ecological niche found by the Chinese had been opened by the near extirpation of sea otters and the subjugation of indigenous people, who had long foraged for abalone, but already, that earlier history was forgotten. By 1850, the Spanish missions that supplanted indigenous villages had themselves fallen into disarray and given way to burgeoning American settlements in what had just become the thirty-first state, California, and the abundance of abalone, with their beautiful shells, seemed to be the natural condition of the coast.[10]

The "abalone rush" began first on the rocky heads of Monterey Peninsula. Word spread quickly through Chinese enclaves in San Francisco, and by the spring of 1853, the *Daily Alta California* reported that five to six hundred Chinese had "almost taken Monterey 'by storm,'" and were engaged in gathering, drying, and packing meat of the "*conchas nácar*," "aulone," or what was commonly called the "California shell."[11] These were mostly intertidal black abalones and red abalones exposed during low tides or that could be gathered from boats in slightly deeper water. With their flat-bottomed sampans, Chinese fishermen could row around rocky headlands to reach remote coves and shallow reefs. While an oarsman deftly maneuvered the boat, a fisher dislodged abalone with a wedge-tipped pole and then gaffed the animals before they fell to the bottom.[12]

Americans had generally ignored the abalone because its unfamiliar meat had a reputation for being odd and tough. By the 1870s, reporters routinely included it in articles that demeaned the "queer" foods of the Chinese, describing it as "tougher than an old boot."[13] One went so far as to characterize the "decidedly uninviting" meat as "dirty saffron color, shot with a sickening red, a more leathery mess of livid looking nastiness it would be difficult to conceive or imagine."[14] But the Chinese used traditional curing techniques to preserve the shellfish, and then savored shavings of its dried, umami-flavored meat in soups and congees. Typically, they removed fresh meat from shells, boiled it in large iron cauldrons for several hours, salted it, and then spread it out on the beach to dry for several weeks. The final result was a hard, wizened chunk the size of a fist that, according to more than one reporter, resembled a "horse's hoof."[15] The fishermen packed the chunks into hundred-pound sacks and shipped them to San Francisco, where they were loaded onto larger boats for export to China. In 1860 alone, 2,391 sacks were exported.[16] As more and more Chinese immigrants came to America to work building railroads, Chinese construction companies diverted considerable quantities of abalone to feed them.[17]

Monterey's abalone rush didn't last long. After the largest, easiest-to-reach shellfish had been collected, most Chinese abalone fishermen traveled south to other rocky hot spots down the coast—San Simeon, Cambria, Port San Luis, Santa Barbara, the Channel Islands, and the Palos Verdes Peninsula. By the 1870s, the front of harvest reached San Diego, which fast became the southern hub of the abalone trade. When Chinese fishermen first arrived, they gathered ample abalone from nearby rocks and reefs, reaching Point Loma and rocky beds up the coast with their sampans and sailboats. After tapping out local areas, they began to sail large junks down the coast of Baja, dropping fishermen at beach camps to gather, process, and dry the black and green abalone that thrived there. Eight companies of abalone fisherman worked the four-hundred-mile shoreline south to Cedros Island in this manner.[18] Before long, Chinese companies built a formidable trade network to accommodate these remote camps. San Diego–based junks would sweep up the Pacific edge gathering dried abalone and hauling it to San Francisco for export. At a time when most boats in America's fledgling West Coast fishing fleet could carry five tons of seafood, Chinese junks routinely hauled fifteen.[19]

Although the Chinese abalone trade focused first on dried meat, there soon developed an international market for the mollusk's unique shells, too. According to malacologist Robert E. C. Stearns, who traveled to California in 1867, San Francisco's Chinese merchants exported 3,713 sacks, or nearly half a million pounds, of abalone shells to China in that year alone.[20] American merchants soon joined the shell business, too, shipping vast numbers to France and Germany, where they were fashioned into stylish buttons and jewelry. Artisans from both sides of the globe prized the unusual iridescent nacre for inlay work in furniture, frames, and even musical instruments. In New York, a dazzling piano was exhibited with "ivories" made of pearl and sharps and flats of green abalone shell. As demand for the glimmering material heightened, the value of the shells would, for a time, surpass that of the meat.[21]

Although few records are available to reliably document the extent of the early Chinese abalone trade, by all accounts, it was tremendous. The *Daily Alta California* reported massive abalone exports, with thousands of hundred-pound sacks of dried meat and shells loaded on ships through the 1860s. By 1879, when the US Fish Commission made the first official survey of California's fisheries, it reported that Chinese abalone fishermen were processing 777,600 pounds of dried meat and 3.8 million pounds of shells per year. The bonanza fishery had removed millions of abalone, mostly from intertidal zones on the mainland, and was starting to tap offshore islands, too.[22]

* * *

The massive buildup of abalone that made the fishery so lucrative to pursue also made its removal by fishermen look like plunder. Not long after the aba-lone rush hit Monterey, local newspapers began to raise concern. As early as 1856, the *Pacific Sentinel* reported that the Chinese had "cleaned nearly all the [abalone] from the waters around Point Pinos." In 1866, the Monterey County assessor complained that the abalone supply was "exhausted."[23] The worry that Chinese fishermen were stripping out abalone became a recurring refrain in communities where locals observed abalone disappearing from intertidal rocks while shells piled up in massive mounds onshore.

When the young fish biologist David Starr Jordan arrived in southern California in 1879, he was highly attuned to such conservation concerns. Hired by the US Fish Commission for the first national survey of Pacific coast fisheries, Jordan's job was to gather information by talking to fishermen, mar-ket men, and newspaper reporters to assess the scale and value of fisheries. The impetus for the forward-looking inventory—and indeed for the US Fish Commission itself, established by Congress in 1870—derived from concern about the plummeting of valuable Atlantic salmon and shad runs. By applying a scientific approach upfront, commission head Spencer Baird aimed to avert such devastating fishery declines on the Pacific coast.[24] Concern about fishes also reflected broader national anxiety about declining waterfowl and other wildlife, which—by the 1880s—had been severely diminished as farms and towns consumed former habitats and as market hunters worked to sate the voracious appetite of America's growing cities. With the vanishing of once-vast buffalo herds, the specter of human-caused extinction raised a new kind of moral worry.[25]

Jordan spent four months traveling up California's coast. No sooner had he arrived than he began to hear complaints about Chinese fishermen. Com-peting Italian fishermen accused them of catching juvenile fishes as bycatch in their fine-mesh shrimp nets, a practice Jordan regarded as "disastrous" for the future of fisheries.[26] According to Jordan, the Chinese fishermen claimed the incoming tides would replenish all the fish that they took and that the ocean was "boundless," following the dominant mercantilist view of marine resources at the time. But having studied the decline of Atlantic fishes under Baird's mentorship, Jordan didn't buy it.[27]

He also became alarmed about California's abalone, the other fishery dominated by Chinese fishers. "By excessive working of this industry," he noted, "the abalones have been nearly exterminated in all accessible places."[28] Jordan regarded Chinese harvests of abalone at the remote Channel Islands as evidence that easier-to-reach mainland shellfish were already depleted.[29] Jor-dan was most critical of Chinese operations south of San Diego, in Baja, where

4.1. A Chinese fisherman and his family set up a stand to sell abalone shells as souvenirs to growing numbers of tourists visiting the Monterey Peninsula. Photo by Joseph K. Oliver, c. 1895–1900 (Pat Hathaway Photo Collection, Monterey, CA, 78-041-0002)

abalone "depredations" were "so extensive as to almost exterminate the species," prompting the Mexican government to set up a special consulate to rein in the fishery. When the US Fish Commission repeated this survey a decade later in 1888, it found the abalone supply in San Diego to be "exhausted."[30]

Jordan's report documented a downward shift from the superabundant abalone populations that everyone had assumed to be natural, but his concerns about "extermination"—garnered largely from talking with white ethnic fishermen—were no doubt amplified by the noxious racial prejudice prevalent in nineteenth-century California. Almost as soon as the Chinese had arrived, they'd met with xenophobic hostility in mining and then railroad camps. In response, they'd set up protective companies to employ and supply fellow immigrants. Along the coast, the trade networks Chinese companies created to market their fish and shellfish minimized interactions with competing fishermen and merchants of different ethnicities. By focusing on a resource niche that everyone else ignored, Chinese abalone fishermen were mostly able to avoid frictions while building a successful export industry.

Nevertheless, racial hostilities in other arenas continued to escalate. Resentful that immigrants often worked for lower wages, labor unions started to rail against the Chinese. When economic depression struck in the 1870s, prejudice exploded in deadly violence in Los Angeles and San Francisco.[31] Even in San Diego, where Chinese fishermen had supplied local markets and gained greater social acceptance, the scene shifted when more competing fishermen and tourists arrived with completion of the California Southern Railroad. Soon, the local newspaper began to carry grievances about "smelly" fish-drying racks around the city's Chinese enclaves. As Asian exports of dried fish from San Diego peaked in 1884–1885, local hotelkeepers complained there were not enough fresh fish to serve their guests and blamed Chinese fishers. Similar antagonisms arose in other coastal communities, too.[32]

As anti-Chinese sentiment reached fever pitch, the US Congress passed a series of exclusion acts in the 1880s that prohibited Chinese laborers from entering the country and eventually barred Chinese junks from entering US waters—effectively ending the abalone export trade. It's difficult to disentangle concern for dwindling coastal abalone from racial animus toward Chinese fishermen, but, ultimately, it was racism that put the kibosh on California's first commercial abalone fishery.[33]

In 1893, the US Fish Commission reported that the Chinese fishery at San Diego had been all but abandoned. William Wilcox, the young fish biologist who followed up on Jordan's inventory, reported that the thirteen junks he'd seen when he arrived in 1888 had been reduced to just one and that intertidal abalone already appeared to be rebounding.[34]

Over the course of several decades, Chinese fishermen had gathered and exported vast numbers of intertidal and shallow-water abalone, making the first incremental dent in the massive surplus of large mollusks that had built up for decades. With the virtual shutdown of the Chinese abalone fishery, California's unique shellfish had survived its first exposure to the limitless demands of the global commodities market and gained a short reprieve until the next group of admirers and exploiters arrived.

ABALONE RISING

With the small bay full of sailboats, boys and girls swimming in the gently lapping waves, the dramatic rise of Sugar Loaf Rock, a delicious breeze, and a local band playing on the wharf, the atmosphere at Avalon could not have been more festive. A steamship had just arrived with hundreds of visitors awaiting the pleasures of vacation. "Catalina is the fad this year," explained the *Los Angeles Times*.[35]

It was 1889, and for about a decade, Catalina Island—a beckoning outlier of rocky land twenty-six miles off southern California's coast—had grown in popularity as a resort. Families and parties of gentlemen hunters and anglers from upscale addresses in Santa Ana, Pasadena, Orange, and Los

Angeles made the two-hour journey by steamship or sloop and then stayed in red-striped wall tents on Avalon Bay. Visitors journeyed out to Catalina to "rusticate."[36] As one explained, "It is the true land of sweet idleness, where one can drift around, with all nature to entertain."[37] With the completion of the Santa Fe Railroad's line to Los Angeles and the construction of the Hotel Metropole in 1887, Catalina soon became renowned for its distinctive brand of southern California vacation.

Visitors not only swam, sunbathed, and danced, but they could also fish for giant barracuda, hunt for feral goats left behind from an ill-fated ranching venture, ride horses on steep trails, row boats in Avalon Bay, and take yacht trips around the island, which included stops on remote beaches or rocky reefs to gather "moonstones" and other "marine curiosities."[38] The island's waters were gin clear and teemed with all manner of life—flying fishes, brightly colored sea anemones, and "moons of the deep" (jellyfishes). But the curiosity that most piqued vacationers' interest was abalone.[39]

Early hikers to the island's interior were fascinated by the large mounds of shells left by Native peoples, which they interpreted as evidence of prehistoric abalone abundance.[40] Treasure-seekers also found artifacts inlaid with abalone. Charles Holder, the renowned naturalist who wrote about the Channel Islands, described an early trip during which he and his guide exhumed a set of bone flutes, decorated with beautiful mosaics of abalone shell.[41] At the time, only some few decried the impropriety of "robbing graves"; for the most part, "Indian relicts" remained coveted souvenirs until Catalina's owners finally outlawed the looting.[42]

Visitors also were fond of gathering "brilliant abalones" at low tide.[43] Holder described their prevalence in his guidebook, *Half Hours with the Lower Animals*: "In some localities, every rock is covered with them, and in places where the black abalone is common, I have found them piled one upon the other."[44] As early as 1889, the island's newspaper recommended an excursion "to the haunts of the abalone" as one of the "vigorous activities" that a vacationer might pursue.[45] Indeed, the *Los Angeles Times* would soon report that gathering abalone had become "one of the most popular amusements indulged in by the enthusiastic tourist."[46] In addition, there was growing interest in studying nature as a wholesome outdoor activity in its own right, with tourists carrying field guides to the tide pools, where they sought out unusual seaweeds, crabs, and seashells.[47]

In the late 1890s, when glass-bottom boats were introduced at Avalon, Catalina's underwater environment became the star attraction.[48] Boatmen took tourists to view "marine gardens" near Sugar Loaf Rock and other fancifully named submarine locales: the "Grand Cañon of the Pacific" and even

4.2. Postcard depicting one of Catalina Island's popular glass-bottom boats. Boatmen dove to retrieve abalone shells for visitors. Photo c. 1903 (From The New York Public Library Digital Collections)

an undersea "Yosemite."[49] Soon, the boatmen started what would become a popular tradition. They dove down into the sea gardens to retrieve abalone shells for their passengers for 25 cents apiece.

If a visitor did not gather his or her own abalone shells or pay a glass-bottom boat man to bring one up, they were available for sale at the Avalon shell shop—au naturel, polished, engraved, or in the form of novel trinkets, such as nacre-inlaid salt and pepper shakers, buttons, or glimmering jewelry.[50] The iridescent shell seemed to hold some of the brilliance of southern California sunshine and some of the sparkle of its clear seawater. And although Californians had become "rather accustomed to overlook abalone jewelry" because it was so common, as one writer explained, "to the Eastern visitor and European tourist, the little trifles made of this shell are looked upon as something . . . distinctly Californian."[51] Abalone shells were fast becoming the quintessential California souvenir.

Part of Catalina's appeal to visitors was the chance to eat fresh seafood and local dishes that could be eaten nowhere else, and so abalone meat found its place on Catalina Island menus. In 1889, the chef at the Hotel Metropole was proudly serving abalone chowder, and in the years that followed, he would add fried abalone and abalone patties to the bill of fare.[52] By 1897, word of abalone chowder had made it east to the New York Times, which described it as "a popular dish in the hotels of the California coast." The mystery of abalone preparation was also already noted as part of its appeal. As the Times explained, "To the uninitiated, its preparation is a mystery as no amount of cooking has any effect upon it until it has been thoroughly pounded with a hammer or a hatchet, breaking the leathery muscle." Only then did the meat become "tender and appetizing."[53] Through time, as southern California resorts began to attract more middle-class vacationers, the mystique of eating abalone would spread.

And so, on Catalina Island, the abalone that had eluded harvest for the Asian export market became the object of a profound cultural transformation. From a "queer" and crude food of the Chinese, abalone morphed into a delicacy of American resort diners and then into a must-try dish for tourists. At the same time, the very work of harvesting intertidal abalone was transformed from a menial job performed by denigrated Chinese fishermen to a vigorous and rarefied leisure activity of highbrow vacationers. As souvenir, as delicacy, as quarry, as subject of nature study, abalone were becoming a valued and distinctive part of southern California's natural scene. On Catalina, abalone became a California icon.

With abalone's growing allure, Catalina tourists gathered "tons of abalone shells" every year, though no records were kept of this early recreational harvest.[54] Abalone hunting became a popular activity on the mainland, too, with guests at the Hotel Redondo's beach camp heading south at low tides to gather hundreds of the shellfish from the rocks of the Palos Verdes Peninsula.[55] It's no surprise that the unrelenting take by enthusiastic tourists soon began to create scarcity.

By 1905, Avalon's glass-bottom boat men had resorted to "planting" fresh shells on the seafloor early each morning to make sure that visitors could experience the abalone abundance they'd come to expect.[56] The *Los Angeles Times* poked tongue-in-cheek fun at this "deep-laid conspiracy" that deceived tourists. "As the tenderfeet don't know abalone shells don't grow empty and bottom-side-up on the floor of the ocean, they are just as well satisfied," the reporter quipped.[57] However, the prank suggested reason for concern. Abalones were indeed becoming harder to find.

AWABI TO ABALONE

After federal exclusion laws ended the Chinese abalone export industry, it didn't take long for the niche of commercial abalone harvesting to be filled once again. By the late 1890s, after several decades' reprieve, Monterey's intertidal and shallow-water abalones were again flourishing. That's when a Japanese woodcutter working for the Southern Pacific Railroad noticed the shellfish clinging to the rocky shoreline. He sent word back to the Japanese government, which dispatched a young fisheries expert to investigate. Seeing the abundance of Monterey's abalone firsthand, the enterprising Gennosuke Kodani leased land on the protected north side of Point Lobos (south of

Monterey) to set up operations, and wired his brother in Japan to hire some divers and come help develop the fishery.[58]

Japan had a long and rich history of abalone diving. There, divers, known as *ama*, were mostly women from villages on the country's crenulated east coast. Holding their breath, they plunged to depths of eighty feet to ferret out the highly coveted shellfish, known in Japanese as *awabi*. Through centuries of diving, the *ama* had developed customary practices and rituals for legendary underwater endurance. They used a distinctive mode of breathing, characterized by a whistling call (known as *isobue*) that enabled them to dive efficiently and to communicate with each other in the water. An *ama* might dive fifty times in an hour, caching her quarry in a floating wooden bucket. Because the Kuroshio current off Japan was typically warm during summer months, the women traditionally dove nude, with only rope belts to hold their pry tools, until Victorian-era modesty compelled them to wear white cotton shirts and skirts. There were male *ama*, too, but they generally speared fish in shallower waters.[59]

Shinto religion, local history, and ancient myth shaped *ama* rituals and beliefs about abalone. According to local lore, when an early princess traveled along Japan's east coast looking for a place to set up a shrine to the sun goddess Amaterasu, an *ama* diver gave her some abalone to eat. Impressed by its delicacy, the princess built the now-renowned Ise Grand Shrine in Shima and asked the *ama* to make offerings of abalone there each year.[60] The women divers have kept up this tradition for more than a thousand years, still ritually harvesting abalone from a sacred cove to supply an annual festival offering to Amaterasu, and also to make *awabi noshi*—stretched, dried strips of abalone meat used to celebrate special occasions.[61]

Ama rituals together with the localized, cooperative governance of Japanese fisheries and an ethic of restraint served to sustain abalone stocks over many centuries. Indeed, the few remaining *ama* divers in Japan today—now elderly—claim their customary practices were always oriented toward "returning the following year."[62]

However, traditional fishing practices began to change in the latter half of the nineteenth century, after US Commodore Matthew Perry forced Japan to open its markets. The long isolationist and feudal Edo period was succeeded by the more market-oriented Meiji government that spurred Japan to catch up with the modern Western world by investing in education, technology, and industrialization.

It was during this modernizing era that Gennosuke Kodani came of age. University-educated with courses in marine fisheries biology, Kodani brought an enterprising approach to bear, born of the new rather than the traditional order.[63] In the 1890s, a series of heat waves, with poor ocean conditions off

4.3. Japanese fishermen first pioneered the use of hard-hat diving gear to harvest red abalone from subtidal reefs near Monterey. Crew members hand-pumped air through a hose to supply a single diver, who passed up baskets of shellfish. Photo by Joseph K. Oliver, c. 1904 (Pat Hathaway Photo Collection, Monterey, CA, 85-020-001)

Japan's coast, had resulted in a massive die-off of kelp and depressed fisheries. With many men out of work, some were willing to travel to California to earn money to support their families.[64]

When the Japanese fishermen first arrived in Monterey, they tried to dive in the traditional *ama* manner, with cotton shorts, shirts, and goggles, but found California's ocean water to be bone-chilling. They persisted, harvesting big red abalone in shallows with glass-bottomed looking-boxes and hooks, much as the Chinese had done decades earlier. At Kodani's request, another crew of divers came from Japan the following year, this time with hard-hat diving suits and air supply tubes.[65] As part of Japanese modernization, male divers had begun to experiment with the new equipment developed first in Europe for underwater construction.[66] The new apparatus enabled divers to take more abalone at greater depths. Traditional female *ama* divers resisted these new technologies out of concern that they'd lead to overharvest, but in California, the promise of technological progress prevailed.[67]

Kodani's crew at Point Lobos pioneered the new diving suits in deeper and colder waters than they'd ever encountered and soon became meticulous masters. Early photos show six to eight men working on a wooden boat to support one diver. An oarsman kept the skiff in place, closely tracking the diver's bubbles. Two crewmembers pumped air down a supply tube with a hand compressor while another made sure the tube never became kinked or entangled in kelp. Another crewman tended the lifeline, attached around the diver's waist, serving not only for emergency escape but also to communicate with the surface. When the diver's mesh basket was filled with several dozen abalone, he gave a yank, and other crewmen hoisted the harvest aboard. Yet another man cut kelp ahead of the boat. Most crewmembers rotated jobs, but the diver worked underwater all day, coming up only for brief breaks.[68] The crew harvested hundreds of abalone daily.

On the beach in what's now called Whalers Cove on the north side of Point Lobos, Kodani's crew built sprawling racks to dry thousands of abalone in preparation for export. In 1898, Kodani entered into a fortuitous business partnership with his Point Lobos landlord, Alexander Allan, and they added a cannery, aiming to also increase domestic sales.[69]

Around the same time that Kodani's divers started work near Monterey, a group of Japanese railroad workers found an abundance of abalone at White's Point on the Palos Verdes Peninsula near Los Angeles.[70] With Chinese fishermen gone for more than a decade, the abalone there, too, had rebounded. The White's Point crew first gathered intertidal black abalone and then worked deeper for green and pink abalone—species prevalent in southern California's warmer waters. Diving with goggles, they used sake barrels anchored as

4.4. Gennosuke Kodani, here with abalone drying racks at Whalers Cove north of Point Lobos, first brought Japanese divers to Monterey, where he started the Point Lobos Canning Company with Alexander Allan. Abalone caught by Japanese divers were exported, but after a 1913 law prohibited sales outside California, Kodani and Allan shifted to supplying a growing domestic market. Photo c. 1904 (Pat Hathaway Photo Collection, Monterey, CA, 2002-076-0001)

floating buoys to serve as both deepwater resting spots and places to cache their harvest bags until a skiff retrieved them.[71]

Before long, Japanese fishing crews had started harvest-and-drying operations at abalone-rich locations up and down the coast, including Mendocino, San Simeon, Cayucos, the Channel Islands, Santa Monica, San Diego, and Baja—many of the same places where Chinese had gathered intertidal and nearshore abalone.[72] Only this time, they took abalone from vast untapped beds in deeper waters with new hard-hat equipment. In 1899, the US Fish Commission estimated that the Japanese fishermen harvested 369,411 pounds of dried abalone meat and 525,453 pounds of shells statewide, about one million animals, for a renewed export fishery.[73]

* * *

Within the first year that Kodani's crew began gathering abalone with their hard-hat dive suits, the local newspaper reported that the Japanese already "had them cleaned off the rocks above the water and for several miles around, and now fish below the surface."[74] Massive piles of shells on the beach were the most striking evidence for the scale of harvest. When local citizens petitioned the Monterey County Board of Supervisors to enact an ordinance to protect the "delicious and valuable" abalone, the *Monterey New Era* vigorously editorialized in favor of restrictions: "At the rate the Japs are operating, there will not be an abalone within many miles of Monterey at the end of the year."[75] Echoing the noxious racial hostilities of earlier times, the newspaper derided the Japanese divers as "cunning Oriental parasites" and contended that the supervisors should not "deprive many poor and deserving people of their own race of a means of livelihood."[76]

Kodani and Allan staunchly opposed the county restrictions. They argued that the Japanese had a long history of harvesting abalone, that abalone multiplied rapidly, and that they were taking only the larger, older animals that would otherwise die of old age and "be lost to the world." [77] Kodani contended that the abalone was "naturally given to us and we should make a profit out of it; if we leave them at the bottom of the sea we are [acting] against the will of God."[78]

In the fall of 1899, the Monterey County Board of Supervisors settled the rancorous local abalone controversy with a compromise—an ordinance that prohibited commercial diving off the Monterey Peninsula but left areas south of Carmel River, including Point Lobos, open to commercial harvest. In addition, it required the Japanese divers to work in waters deeper than twenty feet and to pay an annual $60 license fee.[79] The compromise permitted Kodani and Allan's operations to continue but reserved the abalone of the Monterey Peninsula for locals and increasing numbers of seashell-gathering tourists.

The racialized conflict in Monterey was just one of many that erupted in the face of the expanding Japanese export fishery. Santa Barbara County enacted an ordinance to prohibit export of abalones outside its limits in 1899, and in 1902 required a license fee of $400 per year from Japanese abalone fishermen working on the Channel Islands. Their rationale, according to the *San Francisco Call*, was to "make the license prohibitive" because abalones were "becoming scarce."[80] In 1908, Orange County prohibited export of all shellfish from its waters.[81]

Meanwhile, agitated Mendocino County citizens pressured their board of supervisors for similar action after a camp of Japanese fishermen was found harvesting 2,300 abalones per day. "The camp has been in operation for about four months, and many old settlers have apprehensions that the abalone crop will soon become extinct," reported a 1903 article in *Overland Monthly*. The Mendocino supervisors required commercial divers to pay a $100-per-quarter license fee, a steep charge that the Japanese divers negotiated down to $25.[82] Eventually, though, the crew was reportedly scared off with cannon shots. Modern-day Japanese American urchin (and former abalone) diver Michael Kitahara says some of these early Japanese divers were brutally killed—a dark story that has long remained hidden.[83]

With so many coastal counties adopting different abalone-protection ordinances, the California legislature was finally compelled to address the contentious shellfish issue on the state level. In 1901, after nearly fifty years with no state-level restrictions, lawmakers enacted a minimum size limit of 15 inches in circumference (roughly 8 inches in length) for abalone, aiming to give animals enough time to mature and reproduce before being plucked out of the population pool.[84]

The following year Kodani and Allan pressed the legislature to roll back the new restrictions, particularly for black abalone, the intertidal species that seldom reaches 8 inches in length. However, according to a 1904 state fish commissioners' report, "Representatives from all the counties along whose shores these fish are found were firmly united in favor of retaining the present law without amendment, and refused to recede one iota from their position"—though, somehow, an accommodation was eventually made.[85] In the next few years, the legislature went further, establishing a closed season in the spring to allow for spawning and adopting different size limits for various species—though at the time, knowledge about the reproductive biology and growth rates of California's abalone remained scant.[86]

Although general principles of fisheries management motivated and informed the new abalone conservation laws, racial bias also remained a major factor in how lawmakers considered restrictions. In 1907, legislators prohibited use of all diving gear statewide in response to allegations that commercial divers removed undersized abalone.[87] A later fish commissioners' report explained, "The real purpose of the law was aimed at the aliens, Japanese and Chinese principally, who were taking them by the ton without regard to size by the aid of diving suits, removing the meat from the shell in the water, bringing it ashore, where it was dried and shipped either to China or Japan."[88] In 1908, after the United States adopted a "gentlemen's agreement" with Japan that ended Japanese immigration, Kodani and Allan appealed to the US government and obtained special permission for a limited number of Japanese divers to continue to come and work for their Monterey company under guest worker visas.[89] In 1909, the California legislature would rescind its blanket prohibition on diving gear but require differential license fees for commercial divers: $2.50 for citizens and $10 for "aliens."[90] The most restrictive law was yet to come.

THE ABALONE EATERS

Just a few years after Kodani and his dive crew came to live and work harvesting abalone at Point Lobos, another small group—a set of celebrated writers and artists—came to live and work just across Carmel Bay's "peacock-blue waters," as Jack London would later describe that swath of sea.[91] At Carmel, the writers found cheap real estate, refuge from Victorian-era constraints, natural beauty, convivial spirits, and the "succulent shellfish" with which they'd later become inextricably linked. In fact, poet George Sterling, the charismatic central figure of the Carmel artists' colony, would ultimately become better known for his abalone pounding than his verse.

Sterling had come west from New York in 1890 after trying out seminary and medical schools. In San Francisco he had settled into a promising, if prosaic, life working for his wealthy uncle in real estate development. The young man's restless spirit was stifled by the business culture, so he took up corresponding with well-known writers and crafting poetry on the side. After he published a popular book in 1903, Sterling's magnetic personality and handsome face put him at the center of a vibrant literary scene. Through the city's renowned Bohemian Club, Sterling befriended Jack London, whose *Call of the Wild* had recently thrust him to national fame, and the two enjoyed all manner of lively roistering. However, Sterling found it increasingly difficult to resist "temptations to folly and luxury" in San Francisco, so he moved to Carmel in the summer of 1905, aspiring to refocus his attentions through a

Walden-like experience: living simply and frugally with his wife, growing potatoes, chopping firewood, foraging for seafood, and writing lyric poetry.[92] In the wake of San Francisco's big 1906 earthquake, other bohemian friends followed Sterling's lead, and in Carmel, they together forged new traditions of creative revelry.

Mary Austin, already well known for her acclaimed book *The Land of Little Rain,* moved into a rented house not far from the Sterlings and took to wearing eccentric gowns. Jimmy Hopper, who would become an accomplished newspaper correspondent and magazine writer, set up house with his young family. Jack London never did live in Carmel, but because he wrote vividly about his visits, he would come to be associated with the place. Other writers, poets, photographers, and painters circulated in and out.

Austin would later write about these halcyon days when the Carmel writers worked in the morning and then sauntered down piney paths to the beach in the afternoon to "make impromptu disposals for the rest of the day."[93] Sterling was the "grand pajandrum" of the crowd, exuding a youthful vigor that became an inspiration to them all.[94] According to Austin, he was prone to "restless impotencies" and loved anything that "whetted his incessant appetite for sensation"—hardy hikes, bracing ocean swims, the pull of the undertow, and foraging for shellfish.[95] Indeed, she described "pounding abalone which had just been strenuously gathered from the rocks" as one of the vigorous acts that, along with alcohol, could serve as Sterling's muse.[96]

According to photographer Arnold Genthe, at the Carmel crowd's beach parties, Sterling and Hopper would strip down to bathing trunks and "dive under the deep water by the cliffs and pry abalones from the rocks." He recalled, "The abalone would then have to undergo an hour's pounding with stones—we all took a hand at it."[97]

It was this relentless pounding that inspired the playful Sterling to invent the ritual that would become legend. To lighten up the tedium, he composed a song and made it a game to add rhymed verses ending with the word "abalone."[98]

In what would officially become the opening verse, abalone were literally the key to attaining the cherished bohemian vision of putting life and art before the crass respectability of material status:

Oh! some folks boast of quail on toast,
Because they think it's tony;
But I'm content to owe my rent
And live on abalone.

4.5. In the early twentieth century, famed bohemian writers of the Carmel colony—George Sterling, Mary Austin, Jack London, and Jimmy Hopper—pounded, cooked, and ate abalone, popularizing the shellfish in California's ascending beach culture. Photo by Arnold Genthe, c. 1910, restored (Wikimedia Commons)

The beach parties, the vigorous foraging, and the rite of abalone pounding in the "sacred grove" of pines behind Sterling's house would all have remained the delightful memories of these few friends if not for their celebrity. In 1907, a reporter sent by the *New York Times* to cover the creative scene emerging at Carmel declared, "These people of the West know how to live and feel!" [99] Three years later, the *Los Angeles Times* featured the Carmel bohemians in a tongue-in-cheek Sunday spread with a headline guaranteed to attract attention: "Hotbed of soulful culture, vortex of erotic erudition; Carmel in California, where author and artist Folk are establishing the most amazing colony on earth." The reporter described how Jack London wrote his "red-blooded yarns" taking breaks between chapters to "skirt the fjords in search of the abalone."[100]

The iconic depiction of the Carmel crowd's creative life—and in particular their penchant for beach parties and abalone—would be most widely broadcast and immortalized by Jack London's novel, *The Valley of the Moon*, serialized in *Cosmopolitan* magazine in 1913. In this story, Marcus Hall, a robust character fashioned after Sterling, meets Billy and Saxon, young travelers styled after London and his wife, at a remote beach south of Carmel and initiates them into the secrets of abalone pounding:

> "Now, listen; I'm going to teach you something," Hall commanded, a large round rock poised in his hand above the abalone meat. "You must never, never pound abalone without singing this song. Nor must you sing this song at any other time. It would be the rankest sacrilege. Abalone is the food of the gods. Its preparation

is a religious function." . . . The stone came down with a thump on
the white meat, and thereafter arose and fell in a sort of tom-tom
accompaniment to the poet's song.

In the story, Hall sings the "tony-quail-on-toast" verse, and then carries on
with several others:

> Oh! Mission Point's a friendly joint
> Where every crab's a crony,
> And true and kind you'll ever find
> The clinging abalone.

> He wanders free beside the sea
> Where 'er the coast is stony;
> He flaps his wings and madly sings—
> The plaintive abalone.

He then invites the young couple to join a gathering of the "Tribe of Abalone
Eaters," where the story continues. While women played ukulele and hula-
danced on the beach, men foraged for mussels and abalones until their sacks
were full of shellfish. Then Hall,

> as high priest, commanded the due and solemn rite of the tribe. At a
> wave of his hand, the many poised stones came down in unison on the
> white meat, and all voices were uplifted in the Hymn to the Abalone. . . .

> Oh! some like ham and some like lamb
> And some like macaroni;
> But bring me in a pail of gin
> And a tub of abalone.

"And so it went," London wrote, "verses new and old, verses without end, all in
glorification of the succulent shellfish of Carmel."[101]

In the pages of *Cosmopolitan*, "The Abalone Song" was bequeathed to
the outside world, carrying with it the joie de vivre of the Carmel crowd.
There, on the Monterey Peninsula, in a repeat of the transformation that had
occurred on Catalina Island twenty years earlier, abalone became emblematic
of the abundance, freedom, and good life that California offered.

Already there were two competing visions for California's abalone. In
one, abalone was a food at the center of a valuable commercial fishery and

export industry—in effect a commodity. In the other, abalone played a cherished and iconic role—as a shellfish whose foraging, pounding, and eating could afford self-sufficiency, vigor, conviviality, and creativity while symbolizing natural abundance. The tension between these two visions for abalone set the contours for an intense conflict that would last for more than a century.

ABALONE UNDER STUDY

While California's legislature had incrementally restricted the state's commercial abalone fishery in response to persistent agitation from coastal constituents, the appointed members of the state Commission of Fisheries began to reconsider the Japanese divers' perspective and, in 1909, decided to reverse a previous recommendation to prohibit commercial diving, based on new reports of abalone beds off the South Coast that could "not be profitably taken in any other way."[102] When the commission initiated a systematic investigation of the state's most important food and game species, they directed Charles Lincoln Edwards, a biology professor at the University of Southern California, to study the increasingly sought-after abalone.[103]

With help from a crew of Japanese divers, Edwards sailed out to San Clemente Island, the southernmost of the Channel Islands off the coast of San Diego, donned a heavy diving suit and hard-hat helmet, and took the plunge, becoming the first American biologist to observe California's deepwater abalone in situ.

On his first try, Edwards nearly perished. Owing to an improperly adjusted weight belt, air accumulated in his "trousers," turning him upside down, and he dropped in a headlong plummet until the dive crew yanked him back to the surface.[104] Undeterred, Edwards tried again and, this time, was mesmerized. "Upon the face of a precipice, large specimens of the green and corrugated [pink] abalones rest," he observed. "The shell of each is covered with a luxuriant growth of algae, hydroids, and tentacled tube worms, which mask the creature from its enemies."[105]

Edwards also observed the intensity of the fishery. "I have seen the diver send the net up, filled with about fifty green and corrugated [pink] abalones, every six or seven minutes. During his shift below the diver gathers from thirty to forty basketfuls, each containing one hundred pounds of meat and shell, or altogether one and one-half to two tons."[106]

For the state to manage a commercial abalone fishery, Edwards knew that fundamental information was needed about the animals' life history. Scientists still knew nothing about the breeding habits or embryologic development of any of the abalone species, and so they had no way to determine the best months for a closed season. The recently enacted size limits—intended to rein in fishing pressure—could be counterproductive, he speculated, if it turned out that the largest animals were, in fact, the most prolific breeders, as scientists had recently determined to be the case with the American lobster.

Back in his lab, Edwards closely observed the abalones' behavior in tanks—how they reached with their tentacles to grasp pieces of kelp, how they contracted in response to the shadow of a hand or to light. He noted with particular interest the animals' primitive senses of sight, smell, and taste. He dissected specimens to characterize their basic anatomy. Along the Venice Breakwater, which the city had designated as a "biological reservation," he started an experiment, nesting hundreds of small black and green abalone among the rocks. To study their natural growth rates, he also set forty into a large, mesh-covered concrete box, suspended at mid-tide height to simulate intertidal conditions.[107]

The 1909 revival of the Japanese export fishery in southern California had rekindled alarm about dwindling abalone populations, with many newspapers reporting on the species' impending "extermination." Based on his research, Edwards concluded, "That this is a real danger and not an idle theory is apparent to anyone familiar with the facts." Of San Clemente Island, he reported, "Virtually the whole of the north side has been denuded, and this is also true of the south end." About Catalina Island, he wrote, "Not more than twenty years ago, the green abalones were so thick that they rested upon one another four or five deep, all over the rocks. After much searching in this locality, I was unable to find a single specimen."[108]

Edwards's pioneering abalone research appeared in the very first *Fish Bulletin* issued by the California Fish and Game Commission (formerly the Board of Fish Commissioners) in 1913. Edwards minced no words: "Under our present laws, the abalone is being exterminated," he warned. "We must enact laws that will do away with the piratical robbery of the sea."[109]

Edwards urged the legislature to declare a closed season for at least two years "in order that the preservation of our abalones may be insured," but he envisioned grander plans for the long term. Edwards concluded that the best way to conserve abalone would be to establish a system of reserves all along the coast, with large reservation districts closed on rotation over several-year periods, giving populations a rest from intense fishing pressures. Within those reservations, he proposed establishing smaller perpetual reserves as breeding centers. Since abalones—unlike fishes—were sedentary animals highly vulnerable to being cleared out, Edwards believed this two-tiered reserve system would solve the already long-standing and contentious abalone problem better than would traditional fishery management by size limits and closed seasons.[110]

Edwards's idea for abalone reserves was visionary, predating by nearly a century modern conservation biologists' push for Marine Protected Areas, but, at the time, the Fish and Game Commission was most interested in expanding the economic value of the state's fisheries and never seriously considered it. The California legislature—likely influenced by broader racialized politics—was more receptive to Edwards's recommendations and, in 1913, enacted a law to prohibit all commercial diving south of Santa Barbara, in effect making the state's entire South Coast a reserve and giving deepwater abalone stocks a decade-long break from fishing pressure.[111]

Later that same year, the legislature went further to curtail the Japanese-dominated commercial abalone industry. With an act to prohibit the export of abalone outside the state of California, lawmakers abruptly terminated the trans-Pacific shellfish export industry. According to the *Los Angeles Times*, the law was "considered by many to be a direct slap at Japanese in the state."[112]

Ultimately, Californians' yearnings for natural plenitude combined with unsavory nativism to accomplish a double feat rare in American politics—limiting global trade and constraining a commercial fishery. It was the only time in California history that an active fishery was so decisively restricted—the result of that universal human tendency to point fingers at people considered to be outsiders.[113] Ironically, the political resolve to protect California's abalone would never again be so strong, until it was nearly too late.

The new law gave California's abalone another reprieve, but a new market niche—supplying domestic abalone eaters—was about to emerge.

THE ABALONE KING

By 1913, poet George Sterling, along with some of the remaining Carmel bohemians and arts patrons—now more firmly ensconced in middle age—had found that they could drive over the peninsula in a car and procure their abalones at a restaurant. Café Ernest sat squarely in downtown Monterey. In the small eatery with white linen tablecloths, the bohemian diners exercised their abalone rite in a more decorous manner—by inscribing the playful verses of "The Abalone Song" into the café's guest book.[114] Their celebrity verses and signatures would later stick to chef Ernest Doelter's abalone fillets as mightily as any abalone holds to a rock.

The German Doelter had come to Monterey not long after the Carmel bohemians. Having worked in series of high-end restaurants and hotels, he'd finally set up his own café, staffed by his wife and five children.[115] Before long, he must have discovered the locals' traditional method of abalone pounding.

Abalone, especially in the form of chowder, had already become a popular specialty dish on restaurant menus in California's coastal towns, much like the clam chowder of New England.[116] Locals in the know had long enjoyed cooking the fresh shellfish in soups and stews, with the earliest published abalone recipe (that I could find, see page 46) appearing in the *Monterey Weekly Herald* in 1874. Surviving Chumash, Ohlone, and Pomo people still gathered abalone along the coast. Californians of Mexican heritage prepared the shellfish in a "tender, juicy" manner through a combination of boiling and broiling. In Los Angeles, fresh abalone was sold at the fish market, dried abalone was strung and sold in Mexican groceries along with strings of chilies, and canneries were promoting a new diced abalone product in cans labeled "Eno-laba" (abalone spelled backwards), albeit with mixed success.[117]

Nevertheless, abalone remained an uncommon local dish at best. In this context, Doelter became intrigued by the Japanese abalone industry, which dried massive amounts of abalone for export. He believed that abalone deserved to be eaten more widely in America, and so in the kitchen at Café Ernest, he experimented. In the same way that Germans tenderized veal for wienerschnitzel, he pounded the abalone, dipped the cutlets into a wash of egg, rolled them in cracker crumbs, and fried them quickly in oil. Eventually, using bigger cuts of abalone, he came up with a way to prepare abalone steaks that appealed to Americans' fondness for meaty fare.[118]

Not long after the Carmel bohemians first inscribed their abalone accolades at Café Ernest, Monterey raised Doelter's liquor license fee, so he decided to relocate his large family to San Francisco, where he worked at the German restaurant Hof Brau. Later that year, when the California legislature prohibited the export of abalone out of state, Doelter recognized an opening. Realizing that the only place a diner could now eat a California abalone was in California, he determined to turn the new law's restriction into a business opportunity.

An unabashed self-promoter, Doelter claimed credit for being "the discoverer of Abalone as a white man's delicacy" and set about popularizing the shellfish.[119] With a knack for puffery in an era when marketing was just finding its legs, Doelter sought to dispel the prevailing association of abalone with Chinese and Japanese immigrants and advertised Hof Brau's abalones in the *San Francisco Chronicle* as "The New Seafood as Now Prepared by Us." He listed nine familiar-sounding menu selections: Abalone Cocktail, Abalone Salad, Abalone Nectar, Cream of Abalone, Abalone Chowder, Abalone Steak Hof Brau, Abalone Spanish, Abalone à la Newburg, and Abalone à l'Americaine.[120] Doelter relied on Kodani and Allan for his supply. The still-mysterious preparation method only added to abalone's mystique as a uniquely California dish.

In 1915, when San Francisco hosted the Panama-Pacific International Exposition, abalone was showcased as one of California's top culinary

4.6. One of California's first celebrity chefs, "Pop Ernest" developed a recipe for preparing abalone fillets and promoted the "aristocrat" shellfish as a favorite seafood from his renowned restaurant in Monterey. Photo c. 1925 (Pat Sands Collection)

attractions. The expo featured a special seafood kitchen and dining room with a separate "California Abalone Department." According to promotions, the room was arranged to give the illusion of a cove on the seashore, and diners were promised the opportunity to sample all sorts of California shellfish, but "the succulent aristocrat peculiar to our coasts, the univalve abalone" was singled out for special attention.[121] The Point Lobos Abalone Cannery also set up a booth, decorated with neatly stacked pyramids of cans. With millions of people touring the exposition grounds, tens of thousands must have circled into the sea-cove dining hall to sample the heavily hyped shellfish.[122]

When America's engagement in World War I made Germany an enemy, Doelter left the faltering Hof Brau, returned to Monterey, dropped his Teutonic name, and went simply by "Pop Ernest."[123] For a while, Pop worked for Allan, brokering abalone and mussels and shipping them on ice to restaurants in San Francisco. Then in 1919 he again opened his own place. At the time, there were no restaurants or shops on Monterey's wharf, but Pop's sons helped renovate an old wood-shingled two-story boathouse with a bank of windows that overlooked the bay, and posted a large sign: "Abalone and Mussels a specialty, Pop Ernest, Chef." Newspaper announcements for the new restaurant's opening heralded Pop as "Abalone King." Historic photos show that Pop had, by then, accreted a certain gravitas. Tall and portly, he sported the white linens of a fine chef and a neatly trimmed white mustache and goatee. A distinctive red fez, reputedly acquired at the expo, also became a fixture atop his balding head. The restaurant featured "filet of abalone," traditional abalone chowder served in abalone-shell bowls (reminiscent of the Chumash bowls, but plugged with lead instead of asphaltum), and abalone "nectar," which Pop was keen to promote as a medicinal tonic for indigestion.[124]

The now-aging Sterling and Hopper, plus a new crop of writers and artists from Carmel, once again made Pop's restaurant their meeting spot, and it soon became a favorite for guests of the exclusive Hotel Del Monte, including the likes of Charlie Chaplin. With more cars on the roads, Monterey grew in popularity as a resort destination, and eating at Pop's became de rigueur for visitors. The old guest book from Café Ernest, inscribed with verses from the bohemian's "Abalone Song," became part of the new scene, linking Pop and his delicious abalone fillets to the romance and nostalgia of the Carmel colony.[125]

In 1920, *Sunset Magazine* featured Pop as the chef who knew "more about the abalone than any other chef in California."[126] The reporter linked Pop's fame not only to his cooking but also to the renown of his patrons. In that California way of fame begetting fame, Pop became renowned as one of the West Coast's first celebrity chefs.

Although abalone had been foraged, eaten, and enjoyed by many coastal Californians since long before Pop Ernest arrived on the scene, his masterful promotion—targeting white, upper-class eaters—boosted its appeal and popularity to an entirely new level. Not only did the new prestige of abalone benefit his business, but it also served to keep alive the mostly Japanese abalone fishing industry in Monterey, which readily shifted gears to supply the rapidly emerging domestic market for fresh, locally sliced and pounded abalone fillets.[127]

Pop Ernest perpetuated the mythology that the Carmel bohemians had grafted onto the act of eating abalone, fundamentally changing California's relationship with this most iconic of mollusks. Pop remade abalone into an all-American and distinctly Californian food.

Though the state's 1913 prohibition on exporting abalone reduced commercial harvest pressures for a while, domestic demand for abalone steaks would soon rise to unprecedented heights.

CHAPTER 5
The Abalone Problem

Not long after biologist Charles Lincoln Edwards warned about the perilous overfishing of southern California's black, pink, and green abalone in 1913, the California Fish and Game Commission directed Dr. Harold Heath, an invertebrate embryologist at Stanford's Hopkins Marine Station, to carry out basic scientific research on the most dominant abalone species of central and northern California. Though exports to Asia had ended in 1913, San Francisco's growing restaurant market meant that red abalones were increasingly targeted by Monterey's dive crews.

Heath's graduate student William Curtner began research with a one-year study of red abalones' growth. At the time, the science of dendrology was pioneering the use of annual growth rings to age trees, and the young Curtner was fast to draw similar conclusions about growth ridges on abalone shells. Biologists would later recognize that shell growth does not necessarily occur regularly every year but varies widely depending on food availability. Nevertheless, Curtner's average growth rates became a cornerstone of abalone management in California that endured for nearly fifty years. He determined that it took six years for a red abalone to reach sexual maturity (at the size of roughly 4 inches) and about twelve years to reach the state's legal harvest size (19 inches in circumference, roughly 7 inches in length).[1] That presumably meant that every female abalone had the chance to spawn for about six years before succumbing to harvest.

This finding packed a wallop when combined with Heath's discoveries about red abalones' vast egg producing potential. Mature females, he determined, released up to two million microscopic eggs each year between February and April, which meant that, over a lifetime, every female could release up to twelve million eggs before being fished out. As an invertebrate biologist, Heath knew that most mollusks produced large numbers of eggs because so few larval offspring survived the hazards of ocean life. Nevertheless, because abalone produced far more eggs, he surmised that abalone abundance was rooted in this super fecundity.[2]

Heath had not published this research in any scientific journal, but it came to light when the state's chief of commercial fisheries, Norman Bishop Scofield, pressed him to issue a public statement.

Scofield had studied with David Starr Jordan at Stanford and then built his career with research oriented toward improving fisheries management. He'd started working for the California Board of Fish Commissioners in 1897, conducting studies of salmon, shrimp, and tuna. Through this work, Scofield became a staunch adherent to the emerging idea that science could guide fisheries management to attain a "maximum sustained yield" without depleting stocks. In 1914, he was tapped to head the new state Fish and Game Commission's Division of Commercial Fisheries. Scofield established close relationships with leading researchers at Stanford and at Scripps Institution of Oceanography; set up the nation's first state-of-the-art fisheries laboratory near Los Angeles, at San Pedro; and established the system for collecting commercial landings data that would become the foundation for California's marine fisheries management throughout the twentieth century.[3] Though Scofield was most interested in and occupied by the state's highest-value fisheries, conflict in the abalone arena would persistently hound him.

As center of the abalone industry, Monterey remained a chronic friction spot. In 1907, the California legislature whittled back the county's 1899 compromise abalone ordinance, reopening the peninsula's Seventeen-Mile Drive to commercial harvest and rekindling local residents' ire about commercial divers decimating their shoreline's shellfish.[4] By 1917, residents had become "so worked up of the account of the destroying of this wonderful resource," as one put it, they requested that the state Fish and Game Commission adopt "suitable protection for preserving the abalones."[5]

As Carmel and Pacific Grove became more popular with vacationers, abalone had become a "star attraction" for tourists, they contended, with "hundreds coming from near and far for the sport of prying these shellfish off the rocks and the feast that followed." Yet disappointed visitors were finding fewer and fewer.[6]

Getting nowhere with the commercially oriented Scofield and the Fish and Game Commission, Carmel residents teamed up with the Monterey Peninsula Sportsmen's Club. Across America, anglers and hunters, who increasingly identified as "sportsmen," were beginning to advocate for conservation as a way to preserve opportunities to fish and hunt, especially as market hunting took an increasingly apparent toll on birds, fish, and wildlife. Monterey sportsmen had started to advocate for new laws "to protect the red abalone" because, as they put it, "professional fishermen are rapidly depleting the available supply."[7]

Headed by influential Pacific Grove department store owner Wilford R. Holman, the club circulated a petition, collected seven hundred signatures from Salinas to Big Sur, and convinced their state assemblyman to carry a bill. Holman explained that he had "taken pleasure in gathering abalones" during

5.1. The sport of recreational abalone hunting started with prying the shellfish off rocks at low tide, as pictured here, south of Carmel. Photo by Jan Josselyn, c. 1916 (Pat Hathaway Photo Collection, Monterey, CA, 71-001-PL-09)

low tides for more than twenty-five years and had personally witnessed the once-abundant shellfish become scarce. He was convinced that commercial divers—still mostly Japanese—were to blame. By early March 1917, the outspoken businessman had generated a wave of popular support for legislation that would establish a bag limit for the shellfish and prohibit outright "the use of diving apparatus" along the entire Central Coast, from Monterey south to the Ventura County line—effectively shutting down the commercial industry.[8]

This was exactly the kind of fisheries management that Scofield deplored—with the state legislature making decisions based on offhand opinions rather than scientific research.[9] Already, the legislative body had adjusted abalone regulations during nearly every one of its odd-year sessions. And so, Scofield urged Professor Heath to write up a brief article to explain his recent scientific findings and to "clear up erroneous ideas concerning the danger of exterminating the abalones."[10]

Heath's article, "The Abalone Question," entered the fray of a debate that had already drawn much ink in Monterey and Pacific Grove newspapers.[11] Because abalone produced so many eggs and had several years to spawn before they could be taken by the commercial fishery, and because there were large populations in waters deeper than any divers could reach, continually reproducing, Heath maintained that the shellfish were not in any danger of depletion. Dismissing concern about abalone scarcity, he wrote that "it would be impossible to materially reduce their number."[12] The article was printed in newspapers around the state.

For Scofield, Heath's propitious findings about red abalone's fecundity in Monterey eclipsed Charles Lincoln Edwards's previous admonitions about the need for abalone conservation and put to rest the Fish and Game Commission's lingering concerns about "extermination." With the size limit imposed by the state in 1901, Scofield believed all California abalone had "ample protection." Heath's science gave him the authority to refute and dismiss sportsmen, who increasingly—up and down the coast—claimed to witness declining numbers of abalone on their local bights of shoreline.[13]

The Monterey Peninsula sportsmen took issue with Heath's claims about abalone behavior, questioning his assertion that abalone "remain practically in the same place during life."[14] The sportsmen believed that abalone grew up in deep waters and then moved into shallows, thus continually replenishing the nearshore supply. As one of them explained, "In the 'good old days' . . . people hunting along the shore would take every red abalone of noticeable size" until there were none left. Then, the next day, "they could come to the same spot and find as many again and sometimes more. . . . These abalones drifted and crawled in with the tide from the great offshore supply."[15]

But commercial divers thought increasing numbers of shore-pickers, "taking every abalone of noticeable size," were to blame for scarcity in the shallows and intertidal zone.[16] Meanwhile, Alexander Allan, owner of the commercial cannery at Point Lobos, was pressed to defend his Japanese dive crews in the face of a persistent stream of racial antipathies and mobilized to counter Wilford Holman's campaign. In an op-ed in the *Monterey Daily Cypress*, Allan described data his divers had collected since 1900; despite continual harvest of two tons of red abalone per year from one large submerged rock near Bixby Creek (fifteen miles south of Carmel), he explained, there was "no scarcity"—clear evidence that sustained commercial harvest was possible.[17]

Allan also referred to Heath's research about abalone fecundity and the protection afforded by state-imposed size limits. "Our divers faithfully respect this law," he wrote, but shore-pickers "stalk daily along the same paths" and didn't give intertidal black abalones a chance to grow to full size and reproduce. "Need anyone wonder that the supply has been diminished on the tide rocks of Pacific Grove and Monterey?" he asked. Though recreational shore-pickers were supposed to abide by size limits, too, the state did nothing to enforce the law. Allan tried to distinguish between the black abalone that lived in the intertidal zones and the subtidal red abalone that his dive crews targeted, but this level of biological detail was generally absent from the public debate about abalone.[18]

The local controversy came to a spirited head when Scofield and the state fish and game commissioners finally came to Monterey to hold a public hearing in 1917. Holman would later recall, "It was a standing-room only meeting with people out on the sidewalk." Holman led the charge, and more than a dozen others testified about the need to conserve abalone.[19]

Then Professor Heath spoke. Estimating that two thousand mollusks were gathered from the shoreline each year, he explained it was the shore collector—not the commercial diver—who was "the chief offender" in causing the disappearance of intertidal and nearshore abalone. Heath again explained the prolific breeding of offshore abalone, which he believed dwelled to depths of 250 feet, and recommended setting aside "an experimental ground," closed to both diving and shore-picking, in order to gain more insight into the habits of the animals.[20]

From Holman's perspective, the raucous local hearing was a great success, but with Scofield's help, the savvy Allan made a political end run. His Point Lobos Canning Company treated fish and game commissioners to an abalone banquet and then took them out in boats to show them what Holman would later bitterly describe as "tens of thousands of abalone shells in the bottom of Carmel Bay, which they claimed had gone to waste, dying of old age and starvation." Holman complained that the commissioners "swallowed this statement hook, line and sinker" when instead "the cannery management . . . had dumped the shells into the Bay" to hide the extent of their take from community members. As Holman saw it, the state Fish and Game Commission was too aligned with commercial industry and too little concerned about protecting a public resource from private exploitation.[21]

However, with the scientific authority of Heath's testimony, community sentiment shifted. Three days later, the Pacific Grove Women's Civic Club proposed to reprise the old compromise: to close the Monterey Peninsula to commercial harvest but leave open all areas south of the Carmel River, including Allan and Kodani's Point Lobos—the same deal made by the Monterey Board of Supervisors nearly twenty years earlier.[22] This provided political cover. Behind the scenes at the legislature, Scofield had instigated a similar alternative bill, and the following week, the local assemblyman pulled Holman's bill and, together with the state senator, instead introduced Scofield's legislation—the version that ultimately passed.[23]

The intensity of Monterey's abalone controversy presaged the passion and politics with which Californians would fight over the shellfish for decades to come. Most important, it set the stage for the state's fishery management agency to construe its chronic problem with abalone primarily as a misguided conflict between sport and commercial users. This narrow framework would persist for decades and ultimately affect the fate of abalone in ways no one could foretell.

ABALONE REBELLION

Though Scofield had stayed mostly behind the scenes in Monterey's abalone skirmish, it impressed him deeply as a case where science had the potential to resolve conflicts. "If the facts were only known," he wrote, "there should be absolutely no conflict between the amateur, or sportsman, and the commercial fisherman."[24] But as the controversy erupted in a loose rebellion up and down the coast through the 1920s and 1930s, threatening to severely constrain the growing commercial industry, Scofield went on the offensive.

When the *San Francisco Chronicle* carried the ominous headline, "Abalone Extinct in 3 years, Forecast," and reported that abalone at Mendocino, Pismo Beach, Santa Barbara, and the Channel Islands were almost all fished out, Scofield issued a derisive rebuttal. "There are a few abalone cranks in California," he told a reporter, "and when they cannot find these univalves on the shore, where they are naturally scarce, they say they are becoming extinct."[25]

In a response that would become something of a mantra for the Fish and Game Commission and its subsidiary agency, Scofield contended that abalones' super-fecundity made it impossible to damage the fishery and argued that scientifically derived size limits were the only management tool needed. He even blasted the legislature's effort to establish recreational bag limits—twenty per day near Monterey and ten per day in southern California—as "criminal," contending that it did "not conserve the supply, but hoard[ed] a large food reserve."[26]

In the *Los Angeles Times*, he declared that southern California's abalone had *already* been "saved from extinction" by state size limit laws and commercial diving prohibitions that had been in place for ten years.[27] According to Scofield, offshore stocks had sufficiently recovered from past fishing to reopen them once again to commercial take, but many southern Californians remained wary and opposed the idea. In a letter to a Commerce Department colleague, the exasperated Scofield railed about the "mistaken idea of the southern California residents" that their green abalones needed protection.[28]

Aiming to placate angry coastal constituents, in 1921 the California legislature raised size limits for commercial divers to 8 inches, giving sportsmen first dibs on any red abalone that reached the lower sport limit of 7 inches.[29]

Nevertheless, a few years later, controversy boiled up again in northern California when a single commercial diver (not Japanese, this time) began to harvest abalone off Mendocino with a new, gasoline-powered boat, causing "considerable indignation," as the *Mendocino Beacon* put it.[30] According to the diver, the red abalones offshore were plentiful, in some places stacked three deep, and so he'd readily gathered fifteen thousand in six weeks. However, the local warden determined that not all were legal size, prompting the Fort Bragg Fish and Game Club to send a sharp letter to the Fish and Game Commission demanding that commercial harvest be curtailed in North Coast waters.

Scofield sent a snide reply. "It is my opinion," he wrote, "that the laws protecting the abalone from commercial extermination are ideal, and that it will be impossible for commercial abalone fishermen to exterminate or seriously deplete the abalone supply." He concluded, "There has been a good deal of hysteria on the subject of abalone conservation. There is absolutely no reason why any region should object to taking abalones commercially."[31]

Scofield's dismissive response riled the Fort Bragg sportsmen into action. They persuaded neighboring Humboldt County to join Mendocino in prohibiting the transport of abalones outside both counties, effectively shutting more than half the North Coast to commercial operations through local ordinances.[32] In 1929, local legislators convinced the state assembly to follow suit, closing coastal waters from Humboldt Bay south through Sonoma County to commercial take. A few years later, when Scofield arrived in Fort Bragg for a public meeting, the Fish and Game Club organized a turnout of three hundred agitated sportsmen to make clear that locals had no interest in commercializing North Coast abalone.[33]

The abalone unrest next spread south, after the Fish and Game Commission reopened commercial fishing on the Central Coast and a Japanese dive boat shipwrecked near Cambria, releasing hundreds of undersized abalone to wash ashore. Riled-up residents there pressed the legislature to close the coast—from the Monterey County line south to Morro Rock—to commercial abalone take.[34] They also pressed for a local ordinance to stop shipment of abalones out of San Luis Obispo County.[35] Neither passed, though eventually, a small bight of coast from San Simeon to Cambria was made an "abalone refuge" to mollify local residents.[36]

From his helm at the Bureau of Commercial Fisheries, Scofield had little time or patience for sportsmen's and local residents' complaints. In 1927, when the legislature created the Division of Fish and Game to take over the administrative and research functions of the Fish and Game Commission, the new agency followed Scofield's commercial orientation, responding to citizens' concerns about declining shoreline abalone by pitching blame at the expanding numbers of recreational gatherers. In the division's 1930 *Fish Bulletin*, state biologist Richard Croker lamented that "at every real low tide, many hundreds of tourists and ranchers can be seen going over every accessible reef and ledge 'with a fine-tooth comb.' State and county authorities are hard pressed to enforce the laws on limits and minimum size, which are so easily broken by thoughtless people."[37]

Although shore-pickers were expected to follow state laws, there was little public education about closed seasons, or size and bag limits, and enforcement fell entirely to county wardens. The Division of Fish and Game took no responsibility. Not until 1931 would the California legislature direct the Fish and Game Commission to require a fishing license for recreational shellfish gathering.[38] Still, with shore-pickers blaming commercial divers for declines, the issue of dwindling shoreline abalone went mostly unaddressed. The Division of Fish and Game wouldn't look more closely at abalone until commercial divers began to complain, too.

A NEW RUSH

As demand for abalone steaks increased after the 1915 Panama-Pacific International Expo, Japanese dive crews based in Monterey weren't the only ones to get in on the new rush. Down the coast, Morro Bay residents had long dug Pismo clams and gathered abalone for personal consumption, but with the rising popularity of the "aristocrat" abalone, they began to harvest shellfish for market, too. A small processing company set up in San Luis Obispo and began to ship fresh, market-ready red abalones by train to San Francisco.

In the early 1920s, young Bill Pierce left his family's hardscrabble mining claim in the nearby mountains and headed to the coast in search of work. While gathering abalones at low tide, he observed Japanese divers working offshore reefs, passing large bags of abalone up to deck crews on sleek boats. With abalone stocks dwindling close to Monterey, the divers had ventured south beyond Big Sur to San Simeon and Cayucos. Pierce realized their methods were more efficient than hauling heavy sacks of shellfish over slippery rocks back to the beach. He befriended some Japanese divers, aiming to learn their techniques, and then drove to Monterey to buy a dive suit. When he couldn't find any Japanese-style gear in his size, he bought an antiquated underwater construction suit with leaden shoes and a heavy helmet. Pierce donned the hefty garb—reputedly 150 pounds—and lumbered clumsily into deeper waters never before accessible to shore-pickers. There he found his own mother lode of untapped abalone.[39]

Pierce enlisted his younger brothers to operate his air pump and haul the shellfish he gathered in a small rowboat. Learning by trial and error, he started diving on reefs north of Cayucos. Eventually, Pierce and several of his eight brothers would buy lightweight Japanese dive gear and larger boats with powerful engines that enabled them to run air compressors to supply oxygen

5.2. Workers at Pierce Brothers abalone processing shop in Morro Bay pose in front of a big pile of shells. Through the 1930s, the Pierce brothers, with just a few other commercial divers, harvested more than ten million pounds of abalone from the Central Coast. Photo c. 1933 (Pat Hathaway Photo Collection, Monterey, CA, 92-032-0001)

underwater and also to travel up the coast, where they found abundant and extensive abalone beds. Beckett's Reef—just north of Piedras Blancas lighthouse—became a favorite spot. No matter how many abalone the brothers harvested, they could return several months later and find that new animals had replaced the ones they'd taken; the supply seemed inexhaustible.[40]

When Bill Pierce first started working from shore, he could gather about fifteen dozen abalone per day.[41] By the early 1930s, after the brothers mastered their new gear and boats, each diver was routinely gathering thirty to sixty dozen a day, though agreeable weather and a prime bed could push their take to one hundred dozen each.[42] Eventually, brother Duke earned the family record: 175 dozen red abalone—2,106 animals—taken in a single day.[43] Because it took so long for the small boats to chug up the coast to the best dive areas, the Morro Bay crews started to stay out for a week at a time.[44] In 1935 alone, the Pierce brothers together with a few other Morro Bay divers—a group of only about ten to twelve men—landed more than one million pounds of abalone.[45]

Meanwhile, brother Dutch, who was prone to seasickness, worked on shore to build up the processing end of the Pierce abalone business. He hired crews that hand-trimmed and pounded the steaks with hefty mallets and then boxed them for shipment to San Francisco. By vertically integrating their business, the Pierce family would go on to dominate California's abalone industry for many years.[46]

These stories about Bill Pierce and his brothers were collected by A. L. "Scrap" Lundy, a diving history buff and author of *The Commercial Abalone Industry: A Pictorial History*—the definitive read for those interested in the history of abalone diving, innovations in diving gear, and commercial divers' perspective.[47] A former commercial diver himself, Lundy was intrigued by the mettle of Depression-era divers who learned through trial by fire, jerry-rigged their own air compressors, and MacGyvered their way out of all manner of sketchy underwater situations.

Once, a crewman accidentally dropped the glass faceplate from Charlie Pierce's dive helmet overboard. The resourceful Pierce whittled some cork from a lifejacket, jammed it into the opening, and then dove forty-five feet to retrieve the glass.[48] Though early divers found ample abalone at depths less than sixty feet, thinning stocks eventually compelled them to probe deeper, and some were afflicted by achy joints, headaches, and numbness—preliminary symptoms of decompression sickness, also known as "the bends," which occurs when divers spend too much time at depth and then rise to the surface too quickly, prompting dissolved gasses in their blood to come out of solution inside the body as bubbles. This dangerous but, at the time, little-understood condition could potentially lead to lung damage, paralysis, and even death. Even more treacherous, if divers went too deep without sufficient air pressure in their dive suits, they could be "squeezed"—in extreme cases with intense water pressure crushing a diver's entire body.[49] Tales of terrifying accidents blended caution and machismo in a menacing mix.

Old-timers also told stories of "green" boat operators slicing air tubes with the motor's propeller, stranding submerged divers with no air supply. Even worse was when a diver got "wrapped"—his air tube entangled in the boat's propeller, leaving no slack to surface. That would be the grim fate of Bill Pierce. He died in 1945 at the age of forty-four, when his inexperienced crew didn't have the wherewithal to rescue him in time.[50]

Beyond their legendary undersea exploits, the Morro Bay divers developed a reputation for carousing on shore. Glen Bickford, who would later become a commercial diver himself, was struck by the extremes of the diving life when he first arrived on the scene in 1936. In a letter to his family back in Iowa, he wrote, "I don't see how any man stands it to go down day after day and week after week, 6 or 8 hours a day under water. . . . After a couple of weeks of good weather, working every day, the divers all get screwy as a bunch of pet coons." Bickford described what, in retrospect, may have been symptoms of recurrent decompression sickness: "They get so they don't hear what is said to them and don't remember the things they say and do."[51]

"Scrap" Lundy's book emphasizes the can-do character of the early commercial divers—a small group of rugged men who worked hard, invented new gear, dove to risky depths, and built a multimillion-dollar industry to supply a favored California seafood. By the start of World War II, they had extracted more than ten million pounds of abalone—prying nearly a quarter of a million animals, one at a time—from submerged rocks off the Central Coast.[52] Yet owing precisely to their determination, they would soon come up against ecological limits.

* * *

Back in Monterey, by the early 1930s, the modest fleet of two Japanese dive crews that worked for Allan and Kodani had expanded to fourteen crews and several companies (both Japanese and white) supplying a wharf full of processors and wholesalers. With diesel engines and larger boats, they harvested more than three million pounds of red abalone each year. Abalone had become California's fifth-ranking fishery, and the ever-increasing commercial take finally began to take a toll on deepwater stocks.

From the time divers had started, small groups of men had, in effect, been mining from a large accumulated bank of red abalone that had spread out beyond their usual protected crevice habitats. Though the plentiful animals continued to reproduce, adding to the stock—especially in spots where the undersea topography, ample kelp, and currents were conducive to reproduction and growth—eventually the slow-growing animals couldn't keep up with the fast-growing harvest. In 1928, the Japanese-owned Pacific Mutual Fish Company warned that production was "50 percent under normal." That same year, it was not a disgruntled shore-picker, but longtime commercial abalone diver Henry Porter who declared, "Monterey County beds are now virtually bare."[53]

The Japanese divers blamed massive landslides caused by construction of Highway 1 along the Big Sur coast for killing kelp and smothering productive abalone beds, though the Division of Fish and Game dismissed their concerns on the basis of one staffer's observation from a high cliff at the road's end during construction that coastal erosion was the norm.[54] In early 1926 and then again in 1931, stormy conditions and the warm waters of El Niño likely contributed cumulative stresses to the nearshore marine environment, but no one yet grasped how the recurrent oceanographic phenomenon damaged kelp and disrupted other conditions abalone needed for reproduction and growth.[55]

Henry Porter had a different explanation for abalone's demise: he blamed divers' careless methods of returning undersized animals. After measuring abalone on dive boats, crewmembers routinely tossed the undersized shell-fish overboard. Instead, Porter thought divers should carefully return the undersized abalone to rocky spots where they could reattach and have a better chance of survival.[56]

With all these factors in play, Monterey was indisputably "losing pres-tige" as California's "Abalone Capital."[57] The Monterey dive fleet's migration south in the 1930s repeated the patterns of earlier abalone seekers—to go where the shellfish were biggest, most plentiful, and easiest to gather. As long as the number of divers remained relatively small, it worked for them to move around, harvesting different areas, and giving abalone stocks time to reproduce, grow, and rebound. But this time, as Monterey's Japanese divers moved south, they came into direct competition with Morro Bay divers. In fact, it was not uncommon for the Japanese crews to find their floating wooden abalone caches, cut and stolen by Morro Bay divers who believed the territory belonged to them.[58] Increasing competition on fishing grounds, with more fishers and more efficient gear, meant fewer abalone.

PROBING MOLLUSCAN MYSTERIES

By the late 1930s, Morro Bay's commercial divers also began to complain about diminishing red abalone. Thomas Delmar Reviea had worked with Bill Pierce but soon became troubled by depleted reefs and the Division of Fish and Game's apparent lack of knowledge and concern. Reviea characterized abalone as "California's most abused and least protected natural resource," and he'd lost patience with the finger-pointing between commercial divers and sportsmen. "Each have been blaming the other for the condition and both are responsible." Given the history of extended and extensive harvest, he wrote, "It seems funny that anyone should expect to find a single one of the shell-fish."[59]

In the spring of 1939, Reviea wrote to the state's new fish and game bi-ologist, Paul Bonnot, urging him to investigate underwater conditions and to start an artificial propagation program. If nothing is done, Reviea warned, "we shall find the state almost bare of abalone within five years."[60]

Hired by Scofield, the Stanford-affiliated Bonnot had at first parroted the agency mantra that existing size-limit laws provided ample abalone protec-tion, but he soon realized that research was, in fact, quite meager.[61] Struck by Reviea's gloomy assessment, Bonnot decided to take a deeper look. He met Reviea and other divers from Morro Bay and queried them about the abun-dance and locations of abalone beds. Although Bonnot initially believed that "the average commercial diver" was "an unreliable source of information," he soon discovered that the hundreds of hours divers spent underwater meant the men had useful observations to impart.[62] Bonnot also talked to Japanese divers from Monterey, noting the many "dead grounds" they reported where kelp and abalone beds had been smothered by dirt from Highway 1 construction. He learned that many prolific beds—long known for reliable production—were all "short" on legal-sized abalone, a sure sign of too much fishing pressure. He plotted all the information he gleaned onto a large map to plan his own dive surveys.

That summer, Bonnot learned to dive from Reviea. Then in the fall, he contracted with Reviea and his crew to help inventory sixteen areas from Monterey to San Miguel Island. They wore hard-hat gear with telephone-type lines for underwater communication. All told, they spent sixteen hours on the bottom, from depths of twenty to one hundred feet down—the most extensive underwater survey to date.[63]

Although Bonnot initially intended to conduct a comprehensive census of abalone stocks, he found it difficult to make precise scientific observations. "Everything appears magnified under water and distances are deceptive," he wrote. "A four-inch abalone seems as large as a wash tub until the shell is spanned by the fingers." And there were the trials of rough seas, murky water, and other undersea life; accidentally kneeling on an urchin could puncture a dive suit with dire effect.[64]

Nevertheless, Bonnot confirmed divers' reports about the scarcity of 8-inch red abalones. At San Miguel Island, he documented the problem of illegal harvest, or "poaching," noting the empty home scars left by abalones "stripped" out even though the remote island was officially closed to fishing. He also finally delineated the lower extent of the beds. Up to that point, divers had mostly worked to depths of fifty feet. With Reviea's help, Bonnot determined that the great majority of red abalone lived only to about sixty feet. Knowledge of this lower limit tempered the idea of a boundless stock of undersea abalone.[65]

Most alarming, Bonnot found a lack of young abalone. In most places, he observed 5- and 6-inch abalones in fair quantity, but no smaller animals. "The absence of the small sizes constitutes a serious condition," he wrote. "As the 6-inch and 7-inch abalones reach 8 inches and are taken by the divers, there will be no younger age classes to replace them." The only place Bonnot found a full size-range of animals was at Pebble Beach Cove on the south side of the Monterey Peninsula—a place where locals had vigorously excluded commercial diving for twenty years.[66]

While fishing out of parents, or broodstock, would have diminished productivity, the widespread lack of young abalone may also have owed in part to massive winter storms that hit the Central Coast in 1937, which would have knocked out kelp—abalone's food—and otherwise contributed to poor conditions for settlement and growth of larval animals.[67] Another key factor was that the juvenile red abalones remain very well hidden in rocky crevices until they reach about 4 inches in size, so Bonnot's survey crew likely missed them. At the time, abalone's early life history remained unknown.

With Bonnot's sober findings tempering Scofield's unmoored optimism, the crisis of diminished abalone stocks finally compelled the state to take a more proactive role in managing the commercial fishery. In 1939, the California legislature gave the Fish and Game Commission new authority to require that professional abalone divers purchase revocable licenses, aiming to address the growing problem of illegal take. In addition, the commission followed through on Monterey diver Henry Porter's insight about careless return of animals and adopted a new requirement that all commercial divers measure abalones underwater and replace them directly on rocks.[68]

However, with Monterey dive boats leapfrogging Morro Bay divers to work waters as far south as Point Conception, Bonnot was becoming convinced that an entirely different system of management was needed. Given the lack of juvenile abalones, he predicted that the steadily rising commercial harvest would soon hit a precipitous drop-off. He believed that a fallow period was needed to allow smaller abalones to grow up and reproduce.

While Bonnot supported tighter restrictions on the Central Coast, a dive on the North Coast had convinced him that ample stocks in closed areas there could be fished in the interim.[69] Harkening back to the idea that Charles Lincoln Edwards had first proposed in 1913, Bonnot and his boss Scofield recommended the strategy of rotating commercial harvest zones around the state to sustain landings.[70]

Though for decades Scofield had trumpeted the view that the state's size limits were sufficient to protect abalone, he finally had to concede that the heavily harvested coastline from Monterey to Santa Barbara had "about reached its limits of production."[71] In the spring of 1941, the legislature considered Scofield and Bonnot's recommendation for rotational management of California's abalone. Though the idea was strongly supported by Central Coast commercial divers who desperately wanted access to fresh abalone beds, North Coast businessmen and sportsmen remained adamant about keeping their shorelines closed to commercial take, and the bill was soundly defeated.[72]

* * *

The overshadowing global hostilities of World War II soon affected California's abalone in entirely unforeseen ways. After the bombing of Pearl Harbor, more than a hundred thousand citizens of Japanese descent were forcibly moved from their homes on the California coast to internment camps. With Monterey's Japanese abalone divers compelled to sell their boats, the local fleet was essentially shut down. Morro Bay's abalone fishery was mostly abandoned, too, when white divers moved to southern California to join the wartime work effort by diving for the industrially important seaweed *Gelidium*, used to make products from lubricants to explosives.[73]

The war gave Central Coast abalone a couple of years of respite from commercial diving pressure—some of the fallow time Bonnot sought. But then, in 1943, in response to the nation's wartime demand for protein, the California Fish and Game Commission opened the entire coast—including long-closed areas along the mainland south of Point Conception and out at the Channel Islands—to commercial abalone take. Divers readily started to work the fresh beds surrounding the remote Channel Islands, bringing in large hauls to processors who set up new shops in Santa Barbara, San Pedro, and Newport Beach. Although the North Coast fishery was closed quickly after the war, commercial abalone diving in sunny southern California continued with a large influx of US Navy and Marine "frogmen" veterans joining a growing dive fleet.[74] As harvests ramped back up, the abalone problem would shift southward.

MEMORIES OF ONE JAPANESE AMERICAN COMMERCIAL DIVER

While researching abalone in Monterey, I tracked down local history expert Tim Thomas, who wrote a book about Pop Ernest and leads waterfront tours. After answering some of my questions, he invited me to a special abalone history presentation the following day at the Clement Monterey hotel on Cannery Row. He'd be speaking, along with former commercial abalone diver Roy Hattori, one of few people still living who could remember what the Central Coast's underwater abalone beds looked like in the late 1930s.[75]

The Clement is an elegant hotel, and the second-floor room appointed for the abalone talk was soon packed with tourists. Thomas began by showing some historic photos of the early abalone fishery, then Mr. Hattori stepped up to share personal stories from when he first started to dive in 1937 through the 1940s, when he was finally able to return underwater after his wartime internment. At only 5 feet, 5 inches tall, Roy had a gentle countenance, but once he began to speak, it was clear that he still possessed an outsized vitality—even at age ninety-two.

He began with the story of his first underwater immersion. Most of Monterey's abalone divers had worked diving in Japan and came to the United States for only a stint, but Roy was born in Monterey—only a few blocks from the Clement, it turns out, back when the canneries dominated Cannery Row.

5.3. Japanese American diver Roy Hattori harvested abalone near Monterey in the late 1930s but during World War II was forced to sell his boat and sent away with his family to an internment camp. He was the first to bring white abalone to the attention of malacologists, who delineated it as a new species. Photo by Rey Ruppel, c. 1940 (Pat Hathaway Photo Collection, Monterey, CA, 91-068-006)

Owing to his father's poor health, responsibility for supporting the family fell to Roy. When he graduated from high school, his father outfitted him with a boat and dive suit and took him to the wharf. But his father had never been diving, so when Roy entered the water for the first time, no one had taught him how to offset the building pressure in his inner ears. As he descended with the heavy suit, he felt sure his eardrums would burst. Eventually, older Japanese divers helped Roy learn the trade. That's how he heard stories of the "old days," when Whalers Cove had been full of abalone. By the time Roy started diving, there were far fewer, but "if you looked hard enough," he explained, "you could always find some." After he mastered the underwater work, he could reliably gather about fifty dozen red abalones per day.

Each time he descended, Roy explained, he would take a few minutes to adjust to the cold temperature, watching the beautiful kelp waving in glinting light. He also described his amazement at finding long pieces of kelp neatly folded up underneath the abalone he took. "This animal has no brain, no hands, no nothing," he marveled, and yet somehow it could accordion-pleat kelp into a tight bundle to fit securely beneath its shell.

Once, working close to a cliff wall, Roy was thrown by a surge against the rock. The thin copper plate of his helmet split open, and he felt a stream of icy water pour into his suit. Remembering advice of older divers, he reached into his bag, grabbed an abalone, and slapped it over the hole. The animal immediately suctioned onto his helmet, sealing the gap so Roy could safely surface. "Abalone can really be life savers," he joked with affection for the creatures that had been his quarry.

In the late 1930s, when the Japanese dive boats traveled south of Point Conception searching for more ample beds, Roy found a different kind of abalone than he'd ever seen before. The smaller mollusk was stout, with creamy white tentacles. He gathered a dozen and brought them back to Monterey, where he gave some shells to local collector Andrew Sorenson. Ultimately, Sorenson would be credited with "discovering" the new species *Haliotis sorenseni,* southern California's white abalone.[76]

Abalone diving occupied Roy and sustained his family and crew for five years. Then Pearl Harbor was bombed. Within a week, the US government expelled all Japanese Americans from Monterey. Roy had to scramble to sell his boat and gear for next to nothing. He and his family were detained in Salinas and then sent by rail to Arkansas—an experience of discrimination that left Roy bitter for years. As he explained, he lost his boat, his livelihood, his dignity, and his "belief that America stood by its words of freedom for all." Nevertheless, with his Japanese language skills, Roy ended up serving in the US Army.

Tim Thomas is proud to say that his hometown Monterey—unlike many other coastal towns—welcomed its Japanese American citizens back after the war.[77] It wasn't until 1946 that Roy had the chance to dive again. He was eager to see what had become of the underwater abalone beds while he was gone. He expected they'd be chock-full of shellfish that had grown large during the Japanese divers' five-year absence. Instead, he found formerly abundant kelp forests gone. A thick layer of mud coated the seafloor that he'd remembered being so beautiful. Walking on the bottom, Roy turned to see billows of mud rising in massive plumes behind him. He blamed the building of Highway 1 for much of the damage. Even as he'd watched the beginning of its construction, he remembered worrying, "If that road gets finished, we'll see the end of the abalone." Seeing the formerly abundant abalone beds in such a desolate state, Roy knew he'd have no future in diving. "It hurt me to see the devastation," he explained, with an inflective emphasis on "hurt" that conveyed deep emotion five decades later.

Roy had also heard that, while the Japanese were gone, Morro Bay divers had come north and stripped out the Monterey divers' traditional beds, a charge that didn't surprise him. During the war, there was only one state fish and game warden responsible for policing the entire coast from Monterey to Santa Barbara. Because the Morro Bay divers had their own processing plant, he thought "they could get away with anything."[78] Roy could never return to commercial abalone diving, but he did continue to dive and spearfish seafood for his family, and would later join Monterey's recreational diving club, the Sea Otters.

When Roy finished talking, I joined the crowd in giving him a rousing applause. I couldn't help but think of the ironies accreted in this single lifetime and place. Born two blocks from the Clement hotel, Roy had delivered abalone steaks to the city's best restaurants, braved years of tough underwater work, and was thrown out of his hometown and incarcerated for his Japanese ancestry, yet he still served in the US Army and then persevered to raise his family. Only after the abalone and other seafoods became scarce had people begun to value the marine life that remained in new ways. Now, with the Monterey Bay Aquarium, restaurants, hotels, and shops girding the scenic waterfront, this old-timer was the toast of town—his path through life intertwining with abalone's enigmatic abundance, scarcity, and enduring mystique.

CHAPTER 6
Too Many Abalone Hunters

Around the same time that young Bill Pierce and his brothers were figuring out how to make a living diving for abalone with heavy hard-hat suits near Morro Bay, another group of young men discovered a different lode of abalone in the warm waters off San Diego. Wearing only swim trunks and goggles, they invented a new sport that would transform southern California: skin diving.

BOTTOM SCRATCHERS

It was the late 1920s, during Prohibition. Glenn Orr was earning his living as a rumrunner but also tended line on a dive boat that harvested the seaweed *Gelidium.* One day, off Point Loma, with nothing else to do because his boss was drunk, Orr decided to take the plunge himself. He nearly died because the boss was too inebriated to keep his air supply line open, but Orr was captivated by the fresh adventure of seeing the underwater world, and astonished by the abundance of abalone and lobster.[1] A few weeks later he purchased a pair of swim goggles and invited some friends to join him hunting for seafood.

Although the shallow knee- to hip-deep waters had been raked clean by shore-pickers, the friends realized that, if they headed out farther with paddle-boards to offshore kelp patches and plunged deeper with goggles, they could find a plentiful array of abalones and lobsters. At the time, Orr was twenty-three years old, and his buddies Jack and Ben were eighteen and nineteen. "We were getting together at each other's houses all the time for seafood dinners, so we decided we should just start a club," Jack Prodanovich would later recall.[2] In 1933, the Bottom Scratchers became the world's first dive club. The group would never have more than nineteen members, but they turned out to be a profoundly influential set, becoming a model for dive clubs throughout California and around the globe.[3]

Right from the start, the friends established some tongue-in-cheek hierarchy. Orr became the "Grand Exalted Walrus" and made up what would become the club's renowned initiation rite. To prove their mettle, prospective members had to dive to thirty feet, capture three abalone on a single breath, bring up a good-sized lobster, and grab a five-foot horn shark by its tail.[4]

These feats were as much a testament to the vast abundance of sea life in the waters off San Diego at the time as they were to the divers' fitness. In a 1983 interview, Ben Stone would recall, "You try to tell people how it was, and they just don't believe you. When we first started the club, everywhere you looked there was abalone."[5]

The goggle outings started out as a way for the guys to feed their families and friends during the Depression, but skin diving soon became their passion. Prodanovich worked nights as a high school custodian so he could spend his mornings underwater. Then, during afternoons, he began tinkering with gear. The swim goggles that Orr had bought made them see double because the plates of glass were not properly aligned, so Prodanovich made them all new goggles by inserting round glass from cheap cosmetic compacts into sections of radiator hose that he cut to fit snugly against their eye sockets. Then he made another upgrade: snorkel masks with a single glass plate so he could wear his eyeglasses underwater.[6]

Through the 1930s and 1940s, the Bottom Scratchers dove mostly for abalone near San Diego at Sunset Cliffs and La Jolla, but Wally Potts, who joined in 1939, remembered one trip he and Prodanovich made up to Aliso Beach, near Laguna, in 1941. As they donned their newfangled masks and recently invented swim fins, a few skeptical fishermen on the beach announced loudly and disparagingly that the area was totally devoid of abalone and fish. Undeterred, the Bottom Scratchers finned out to where it was deep. "We saw big green abalone all over the place," recalled Potts. "We just reached down and grabbed enough for dinner. I speared a few fish on the way in, and when those fishermen saw us, well, their attitude was completely different." By the time Potts and Prodanovich left three days later, the Aliso Beach fishermen had purchased their own spears, fins, and goggles. As word got out, the sport of skin diving quickly grew in popularity.[7]

The Bottom Scratchers catapulted to fame after Lamar Boren joined their ranks. A pioneering underwater photographer and filmmaker, Boren's photos were featured in a 1949 *National Geographic* article about the group's "goggle" diving and spearfishing for giant-sized fish.[8] Boren then filmed a pilot segment for what would become the immensely popular, late-1950s television series *Sea Hunt,* using the Bottom Scratchers as his models. Prodanovich ultimately made much of the dive equipment for the show.[9] In this unwitting way, the dive club played a key role in publicizing the thrill of underwater recreation.

With the advent of the Aqua-Lung air tank by French diver Jacques Cousteau and engineer Émile Gagnan, and its subsequent import to Rene's Sporting Goods shop in Los Angeles in 1949, the sport of diving truly took off. Through the 1950s, American divers began to experiment with the

6.1. The La Jolla–based Bottom Scratchers launched the sport of skin diving and became a model for dive clubs worldwide. Founding members Jack Prodanovich and Wally Potts are shown here with a haul of abalone on their dive board. (Photo courtesy Terry Maas, BlueWater Freedivers)

"self-contained underwater breathing apparatus" that would become known as SCUBA, and new rubber diving suits.[10] While breath-hold skin diving had required a high level of fitness and mental determination, scuba diving could be enjoyed by far more people.

In the postwar era, with millions of new people moving to California to take advantage of the state's economic opportunities, the number of divers skyrocketed. By the mid-1950s, there were thousands of divers and more than 130 dive clubs in southern California, with colorful names like the Neptunes, the Sea Downers, the Palos Verdes Sea-Horses, and the Laguna Sons of Beaches.[11] Many divers subscribed to a new magazine, *Skin Diver*, to share info about club activities, new gear, fishing techniques, and even recipes for preparing the unfamiliar wild seafoods, including abalone.[12] California's trendy *Sunset Magazine* also began to feature recipes for abalone stews, chowders, and steaks.[13]

With the fast-rising popularity of scuba gear, the new recreational divers gathered untold numbers of abalone, lobsters, and fishes from southern California's mainland subtidal rocks and reefs. Before long, they, too, began to confront diminishing stocks.

The Bottom Scratchers first became concerned about the future of the sea's abundance through a connection to Scripps Institution of Oceanography, made with the induction of two Scripps divers, Conrad Limbaugh and Jimmy Stewart. Limbaugh and Stewart introduced fellow members to the renowned ichthyologist Carl Hubbs, who admonished that the giant fishes they spear-hunted in La Jolla Bay represented an isolated and finite group of broomtail groupers. Although the Bottom Scratchers had already pressed the state legislature to sanction spearfishing as a legitimate sport in California, club members immediately stopped hunting the mammoth fish and soon declared use of the specialized guns "unsportsmanlike" and appropriate only for self-defense.[14] By 1951, the club had officially adopted a conservation ethic, including a pledge "to prevent waste of catches."

They also weighed in on abalone. First, the Bottom Scratchers advocated lowering the sport bag limit from ten to five abalones per day, a measure the Fish and Game Commission readily adopted in 1949. Then, the club pressed to ban use of scuba tanks for take of lobsters and shellfish in shallow nearshore waters.[15] As Potts later explained, "The poor old abalone, he doesn't run very fast. When you're skin diving, you couldn't take them all if you wanted to, but a scuba diver can go down and clean a place out."[16] In 1952, the Fish and Game Commission adopted the protective ban.[17]

However, not all dive clubs agreed, especially since the number of permitted commercial divers had more than quadrupled during the previous decade to three hundred.[18] As Woody Dimel of the Los Angeles Neptunes put it, "We are all for conservation, but why should the skin divers have to conserve to fill the commercial man's pocketbook?" In *Skin Diver* magazine, Dimel explained that "the average commercial 'ab' diver gathers from 50 to 60 *dozen* abalone per day"—equivalent to the bag limits of more than 120 skin divers. "Multiply this by the number of commercial divers and you have a staggering total."[19] Moreover, not all commercial divers stayed out of shallow waters, amplifying frictions between the groups.[20]

That year, eleven dive clubs, including the Neptunes, founded the Council of Dive Clubs to represent their recreational interests at Fish and Game Commission meetings. The following year, in response to the council's pressure, the Fish and Game Commission revoked the ban it had adopted, once again allowing sport divers to use scuba gear to take abalone in shallow waters along the coasts of central and southern California.[21]

Yet still recognizing the need for conservation, sport divers aimed to cut back the massive abalone take of commercial divers, which in 1951 surpassed four million pounds for the first time. Flexing its new muscle, the dive club council pressed the legislature to close the mainland coast south of Santa Barbara to commercial abalone diving for two years, pending the results of a state survey to evaluate the feasibility of opening the entire California coast to commercial harvest on a rotating basis. The closure bill readily passed the state senate but was narrowly defeated in the Assembly Fish and Game Committee by a tie vote.[22]

In 1954, Bill Barada, a recreational diver who'd pioneered rubber wet-suits and became president of the Council of Diving Clubs, identified what was already becoming an entrenched political dynamic. Sportfishers were quick to welcome restrictions at the first sign of depletion and could be readily constrained through the regulatory power of the Fish and Game Commission, but the commercial industry was regulated primarily by the state legislature, where, he explained, it was "almost impossible to pass any realistic conservation measures." Though, in the past, the legislature had been responsive to the conservation concerns of local citizens and recreationists, by the 1950s, as the commercial abalone industry rose in economic importance—and, frankly, became more white—the legislature was increasingly inclined to defend industry interests. "The hue and cry over taking away means of making a living . . . has always managed to block even the slightest attempt at control," Barada further asserted. "The fact that over-fishing will put everyone permanently out of business" seemed to be perennially overlooked.[23] The intractable dynamic would persist.

The recreational divers had made their new political might known, but the fishing pressures exerted by their increasingly popular sport would continue to compound the significant pressures exerted by growing numbers of commercial divers. Facing divers with both scuba and surface-air-supply gear, abalone would have fewer areas of refuge.

Ultimately, the Bottom Scratchers would be compelled to change their renowned initiation rite. Abalone became so scarce in favorite nearshore areas that a southern California skin diver would have drowned long before being able to find three on a single breath-hold trip to the bottom.[24]

TRACKING ABALONE

Following World War II, California's Division of Fish and Game (CDFG) began to more closely track the expanding commercial abalone harvest. With the massive postwar economic boom came a threefold increase in the number of commercial divers, and the CDFG reported increasing incidents of illegal take of undersized or short abalone as new divers with little experience struggled to make a living. In 1950, the division reported that "the drain on District 18"— roughly the Central Coast from a bit south of Point Lobos through Ventura— "was excessive and practically all the legal sized abalone were removed."[25]

The CDFG finally recognized the impact of growing numbers of sport divers on abalone stocks, too. Wardens began keeping informal tallies on the North Coast, where recreational gathering during low tides of spring had grown exponentially after the opening of the Golden Gate Bridge in 1937. As one travel writer noted, the "unique sport . . . peculiar to the California coast line," brought out "predawn motorcades that carry hundreds of those seeking the thrill of abalone fishing."[26] During one three-day low-tide period in 1949, the wardens found that 3,902 sportfishers took 15,514 abalones from the popular forty-mile bight of coastline between Fort Ross and Point Arena.[27] There was no commensurate survey of the fast-growing recreational harvest in central and southern California, but the CDFG estimated the sport catch of abalone in 1951 to be "more than half the commercial catch."[28]

Although oversight of the abalone fishery had long fallen under the purview of the California legislature, with increasing concern about declining stocks and continual skirmishes between commercial interests and sport-fishermen, legislators had finally delegated authority to the Fish and Game Commission to oversee some elements of the commercial fishery and to regulate the sportfishery entirely.[29]

Yet one contentious, unresolved issue remained squarely in the legislature's province: rotating the commercial harvest statewide. With declining availability of abalone on California's Central and South Coasts, commercial abalone divers pressed harder to open the North Coast to commercial take. Once again, North Coast resort owners, businesses, and communities staunchly resisted, arguing that a commercial dive fleet would destroy their abalone stocks and sportfishery, which had become important to the local economy.

In 1951, tired of the decades-long squabble between commercial and sportfishermen, the legislature directed the Department of Fish and Game, recently elevated from division status, to start a scientific research program to study and clarify—once and for all—the effects of the commercial abalone fishery on the nearshore sportfishery. The agency still knew too little about the biology of the creature everyone sought to fish. The state's new abalone program was tasked with investigating how stocks were standing up to increased fishing pressure—both sport and commercial—and also to consider the feasibility of opening the North Coast to commercial divers.[30] The political need to referee the commercial versus sportfishing conflict drove the state's science agenda.

The chief of the department's Bureau of Marine Fisheries directed staff to convert an old patrol boat into a research vessel and put Paul Bonnot in charge. Although Bonnot was already convinced that the offshore fishery had no effect on nearshore abalone, he lacked scientific evidence. This time, he intended to tag abalone and track their movements. Bonnot also planned to conduct the first rigorous survey of the North Coast's red abalone stocks to determine whether or not the reefs there could withstand commercial harvest.

Straightaway, the new program hit a series of major setbacks. One month into the research, Bonnot died of a heart attack. Former commercial diver and project skipper Thomas Reviea carried forth the tagging, but not until the following year did the department hire a new biologist, Keith Cox, and another former commercial diver, Glen Bickford, to replace Reviea.[31] Like most marine biologists of the day, Cox had first to learn how to dive, so the veteran Bickford showed him the ropes.[32] The new crew soon found that tags placed during the program's first season had badly corroded, and they had to devise a new system using stainless steel. Then, at the end of the second season, the program's "mother ship" was lost to a storm, leaving the crew without a reliable dive boat.[33]

Despite the setbacks, each year the crew worked from south to north, tagging abalone and aiming to assess stocks. Eventually, Cox and crew obtained Bickford's old abalone dive boat, converted it into a research vessel—the R/V *Mollusk*—and would accrue more than 1,500 hours of underwater observation through the early 1950s.[34]

In a preliminary report to the impatient legislature in 1955, Cox described some early discoveries. He found that the hitherto rarely observed, juvenile red abalone actually hid in crevices or under rocks for their first four years of life—assuaging some of the concerns previously raised by Bonnot about the lack of small, upcoming abalone.[35] Cox also determined that red abalone spawned in spring and summer, not in winter as had long been the basis for the state's closed season.[36]

It took more time to gather the data needed to answer the bigger questions, but after four years of tagging studies, Cox believed that he'd finally resolved the decades-long dispute between commercial and sport divers about the movement of red abalone from deep to shallow waters. The state dive crew had tagged and released more than four thousand abalone at varying offshore depths along different parts of the coast. When they checked the tagged animals one year later, most had remained in the same places they had been released.[37] Later researchers would find that the North Coast's red abalone actually did move seasonally into shallow waters to feed on washed-in kelp, but based on his research, Cox concluded that there was "little, if any, basis for clashes of interest between the sportsmen and the commercial divers."[38]

Most important were stock assessments. Because a comprehensive inventory of the North Coast's abalone proved impossible, the survey crew ended up conducting a more qualitative evaluation, relying heavily on their commercial diving experience and "knowledge of how productive commercial beds 'looked.'" Although the crew found large concentrations of red abalone in some spots, they determined there were not enough for sustained commercial harvest—especially because they also found that abalone growth on the North Coast could be extremely slow.[39]

As a result, in a widely publicized 1957 report to the legislature, the De-partment of Fish and Game strongly recommended against opening the North Coast to commercial harvest—a sharp turnaround from Bonnot's earlier view. Cox explained that North Coast's slow-growing abalone had dark meat that would fail to meet commercial standards; plus, frequent rough weather, heavy wave surges, and lack of safe harborage made it unfeasible to run commercial operations. Finally, because the seafloor dropped off steeply on the North Coast, abalone clustered close against the shoreline with no offshore beds—a configuration that would make it nearly impossible to demarcate separate zones for sport and commercial fishing. And if commercial divers interfered with sport activity near shore, Cox warned, it would inevitably lead to "ex-treme friction"—recalling that when the North Coast was temporarily opened to commercial harvest during wartime, local vigilantes had shot at intruding dive boats. North Coast legislators also voiced opposition.[40]

Cox had carefully considered the idea of rotating harvests statewide to provide for the commercial abalone industry and found it unfeasible; but if not that, then what?

* * *

Looking back, it wasn't just size limits and closed seasons that had sustained the state's rising abalone harvest. Commercial exploitation had already moved from place to place as divers left behind picked-over beds to find more abun-dant reefs. Heavy fishing of red abalone near Monterey had hit its limit around 1929, to be followed by a period of continuous take of red abalone on the Central Coast through 1941. Then, after the war, when southern California and the Channel Islands were reopened to commercial fishing, many divers started to work fresh beds of red and pink abalone there. In effect, the harvest had already been "rotated."

As Cox aimed to make sense of Bonnot's earlier findings and of what he heard from commercial divers, what he saw underwater, and what the new data was revealing, he began to discern larger trends. For red abalone, the harvest had exceeded three million pounds per year in the late 1920s and early 1930s but, by the mid-1950s, was leveling off at about 1.5 million pounds in spite of increasing numbers of divers. Cox realized that the growing take of pinks in southern California and the Channel Islands would inevitably level off, too, with no more fresh beds to tap.[41]

Future production would have to come solely from the annual growth of up-and-coming animals, but Cox's realization that overall landings would soon need to stabilize at a lower level was at odds with the ever-growing number of commercial and sport divers, expectations for still-rising harvests, and the department's long-standing laissez-faire approach to the fishery, anchored to the belief—going back to Scofield—that size limits and closed seasons could adequately provide for a continuing supply of the favored shellfish.

As the abalone program's work on the North Coast wrapped up, new concerns demanded Cox's attention in southern California, where the commercial catch of pink abalone was dropping and legal-sized animals were becoming scarce. In 1955, recreational dive clubs returned to the capitol and this time successfully pressed the legislature to close southern California's mainland coast to commercial abalone diving for two years to give the department time to determine whether the area could continue supporting both commercial and recreational take. Cox started new tagging studies to obtain growth data specifically for pink abalone.[42]

Cox also conducted a series of abalone-transplant experiments to determine whether the technique could be used to "build up" diminished populations. The most important of these aimed also to gauge the effect of pollution—increasingly prevalent in southern California's once-clear waters. The crew transplanted sickly "shrunken" black abalone from White's Point, where Los Angeles County's sewer outfall now flushed "millions of gallons of pollutants" into the sea, to the clean waters off Catalina Island—and conversely moved healthy black abs from Catalina to White's Point. Just one month later, Cox found that all the abalone moved away from White's Point had gained weight, while only half the abs moved to White's Point had managed to grow just a sliver; the others were entirely stunted.[43] The findings suggested that urban pollution had the potential to significantly affect abalone health and growth, but the problem fell beyond the department's purview and led to no further inquiry at the time. (It was not yet known that the White's Point outfall had been discharging DDT from the Montrose Chemical plant into the ocean for more than a decade, a problem that wouldn't be curtailed until 1970.[44])

* * *

Despite continuing complaints about too few abalone—with ten times more commercial divers than there had been twenty years earlier—the state's overall commercial abalone harvest, including pinks and reds from southern and central California, peaked in 1957 at an all-time high of more than five million pounds, about one-and-a-quarter million animals.[45]

However, that fall, conditions for abalone markedly deteriorated with an influx of warm water that persisted through 1958 and 1959, killing off much of southern California's kelp, leaving many of the mollusks to starve. Rather than respond with a protective measure, the Fish and Game Commission lowered the legal-size limit for red abs from 8 inches (in effect since 1921) to 7 ¾ inches, so that commercial divers could find enough legal-sized shellfish to continue working. According to former commercial diver Buzz Owen, Cox had initially wanted to lower the size limit to 7 inches, but many divers objected, fearing that far too many animals would be cleared out. Still, the night before the new size limit took effect, divers staged up near "short beds" to take advantage of the change. Cox later reported that the "expected benefits" of the measure were cancelled out because of the reduced amount of meat from the shriveled animals.[46] His tagging studies showed that the animals' "growth practically ceased and body tissues appeared to shrink," but at the time, there was not much consideration that continued fishing in the wake of such a substantial kelp die-off could harm California's long-abundant abalone.[47]

Cox warned that the "intensive harvest of 1957 and the lack of an adequate food supply" would lead to lower catches.[48] And though landings did decline from the high peak, huge harvests of abalone continued unabated in southern and central California—with an average of more than 4.2 million pounds taken by the commercial fishery each year for the next ten years.[49]

Not until decades later would marine biologists recognize how El Niño conditions could knock out spawning and successful reproduction for several years in a row, effectively precluding the recruitment of whole age classes of abalone. With continued heavy fishing on up-and-coming animals as they reached the new 7 ¾ inch legal size, the missing age classes wouldn't be detected until many years later, when the next generations of shellfish failed to appear.[50] The many-years lag between cause and effect made it impossible to foresee the approach of a critical tipping point.

In fact, when cool waters returned, the quick rebound of giant kelp made abalone in central California and around the Channel Islands start growing so fast as to create a confounding balloon of supply. In 1962, when many red abalone reached the new, smaller, legal-size limit, divers landed so many that Morro Bay processors couldn't keep up and had to turn away shellfish at the dock for the first time in the fishery's history. One local resident recalled fishermen driving around town—pickup trucks filled with abalone—trying to hawk the surplus shellfish to anyone who would take them.[51]

At the time, abalone seemed so abundant no one could imagine that they would ever be anything but plentiful.

* * *

Bob Kirby would later remember legal-sized abalone in southern California "thinning out" around that time. Kirby became renowned among divers for developing the specialized helmets used for deep-sea diving in the Santa Barbara oil fields and for undersea construction around the world, but back when he was starting out as a commercial abalone diver, he found it hard to make a living. Old-timers' stories about Beckett's Reef, with "so many shellfish a blindman could get a load," drew Kirby and fellow divers north to the Central Coast.[52]

His first day there, he remembered, the water was so clear that, floating on the surface, he could see abalone everywhere. He thought he'd be "rich by the end of the day." But once he started measuring, he found almost all of them were shorts. "We were faced with this fascinating and frustrating trance all during the three months we worked Beckett's Reef," he'd later write in his autobiography. Each day he worked twelve hours to harvest fifteen dozen legal-sized abalone, a far cry from the sixty to hundred dozen per day that the Pierce brothers had readily collected twenty-five years earlier. All the time, he hoped to find "one small virgin area," but none existed. "They had all been worked over by other divers, eager as I to make a living," he recalled.

Eventually, Kirby and his buddies cruised down to check out Purisima Point, south of Avila. With little luck there, they decided to scout a kelp patch farther offshore, and finally found what they'd been hoping for—a "virgin bed." With huge abalone—well beyond the size limit—everywhere, they filled up bag after bag, loading the decks of their boats to overflowing.

For the young abalone divers, it was a high point—a sliver of experience that connected them to commercial diving's heyday—and, finally, a lucrative payday. Kirby would later write, "In a half hour, I had picked the virgin spot clean. Then I made a wide circle. There were no abalone left."[53]

✳ ✳ ✳

Before Department of Fish and Game biologist Keith Cox could conclude his research, analyze his data, and make recommendations to improve management of abalone in central and southern California, the legislature cut the state abalone program in 1960 from a full-time investigation to just a two-month annual survey.[54] With rising fishing pressures on the highly valued shellfish, the timing couldn't have been worse to slash funding for abalone research.

Cox continued with cursory monitoring and wrote up what would be his most enduring contribution, a CDFG *Fish Bulletin* devoted entirely to abalone. Intended for both a scientific and public audience, the booklet, *California Abalones: Family Haliotidae,* drew on history and biology to assemble all that was known about the shellfish at the time. In it, Cox shared some important recent findings: abalone didn't grow as quickly or predictably as had long been assumed; kelp die-offs could significantly affect the animals' ability to survive, grow, and reproduce; pollution could impair abalone's health and growth; and the number of divers—both commercial and sport—was skyrocketing to record highs. And because few people could distinguish among the various abalone species, he included photos to highlight the differences in their shells.[55]

Cox had identified a whirl of problems that would soon bring California's abalone century to an end, but persistent and high overall landings—the pulse by which the CDFG measured the success and promise of its fisheries—obfuscated the warning signs. It's hard to know what Cox truly thought, but in its 1962 official *Biennial Report*, the department rosily reported, "From data and observations collected over the last ten years, and in particular, the last biennium, it is evident the abalone resource is in a healthy condition. Present laws are adequate to protect the resource, and barring unforeseen natural phenomenon, California abalone will continue to support a unique fishery for the pleasure and profit of the people of this State."[56]

Biologist Keith Cox had aimed to resolve the abalone problem he inherited from his predecessors—a conflict between recreational and commercial divers. But changing environmental conditions and an entirely new and unexpected competitor were already starting to make the problems of abalone far more complicated.

PART 3

Scarcity
(1939–1985)

In the nine mile coastal area immediately south of the sea otters present range, fishermen harvest nearly one million pounds of abalone annually. It is only natural for sea otters to migrate into this lush feeding area.

> —Earl Ebert, *Transactions of the Western Section
> of the Wildlife Society* (1968)

What we are faced with here, gentlemen, is the same old type of situation that has existed between those who enjoy the aesthetic values of our animals and creatures and those who enjoy and eat seafood or meat.

> —John Gilchrist, California Seafood Institute, Senate
> Fact Finding Hearing on Sea Otters, 1963

As has been true with practically every wild animal in the vicinity of man, the otter is thought to compete with an economic value that man claims as his own.

> —Margaret Owings, *Monterey Peninsula Herald*, March 11, 1968

There are huge mounds of abalone shells in sight of the highway north of Santa Barbara—not put there by the otters. [They] need shellfish to survive, we don't. Frankly, I'm fond of the succulent abalone but have refrained from eating it for years.

> —Mrs. Henry T. Read, *Otter Raft* (newsletter), 1973

There was this sense that abalone were never going to disappear and we could just keep diving for them.

—Ken Nielsen, retired commercial fisherman
and current abalone science volunteer,
"The Era of Abalone-Rich Waters," *NOAA Fisheries Stories*, 2015

I have often wondered how anyone of any vision could look at the landing data alone and say nothing was needed to help the declining abalone stocks.

—Richard Burge, former CDFG abalone biologist,
interview, August 2019

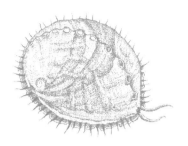

CHAPTER 7
The Sea Otters' Return

Preparing for another slow spring day at her teahouse perched above the ocean on the remote road to Big Sur, Mrs. Frida Sharpe paused for a moment to take in the view. The year was 1938. Through the lens of her spotting scope, she scanned the water and was surprised to spot a group of furry brown animals at the base of rugged cliffs. Realizing that they might be sea otters, she and her husband promptly called the fish and game warden stationed at Monterey. After all, sea otters hadn't been officially seen off California's coast for more than twenty-five years. The last known otter had reportedly been shot near Fort Ord, on Monterey Bay, in 1911, the very same year that an International Fur Seal Treaty had finally conferred some measure of protection for the fur-bearing animals.[1] With no less irony, the State of California took action two years later, passing a law to protect sea otters.[2]

The warden was skeptical, thinking it far more likely that the Sharpes had seen the bobbing heads of bull kelps or harbor seals, either of which, to an untutored eye, could resemble otters. At that point, few living Californians had ever laid eyes on a sea otter. Yet Mrs. Sharpe insisted, and so the warden invited Stanford zoologist Dr. Harold Heath—the same professor who had extolled abalone fecundity two decades earlier—emeritus, at this point—to join him for the drive fourteen miles south of Carmel to check out her report.[3]

There near the mouth of Bixby Creek, the warden and zoologist observed a "herd" (now called a "raft") of about ninety-six sea otters, and the animals long thought to be extinct were officially "rediscovered" in California.[4] Big Sur ranch families would later admit to knowing about the otters' presence but keeping it secret to protect them.

The opening of the new coast highway in 1937 made the sea otters' rediscovery inevitable and increased the risk to their safety. At the time, the monetary value of an otter pelt was about seven times higher than the fine for killing an otter.[5] The Division of Fish and Game immediately positioned a patrol boat nearby to protect the otters from poachers and, shortly thereafter, began conducting surveys from the highway. The division soon counted about three hundred hitherto-undetected animals, and in 1941, the state established a new "sea otter range" along the coast south of Monterey as a refuge.

During the animals' twenty-five-year Big Sur exile, much about America had changed. The sea otters were now welcomed back not for their fur but as wildlife to be appreciated. Smart and playful, the otters fit their new role well, enchanting tourists who watched with binoculars. For many, sea otters embodied the hopeful prospect of bringing an animal back from near extinction and, even more, the possibility of a redemptive turning point for a culture that had destroyed the great bison herds and extinguished the once-abundant passenger pigeon.

Although zoologists who observed the sea otters noted right away their penchant for munching abalones, there was little inkling of trouble. Though it was later learned that Japanese abalone divers had occasionally encountered otters, the animals tended to quickly flee from the fishing boats. Then, wartime internment had forcibly removed Japanese divers from fishing grounds south of Monterey that overlapped with the sea otters' remote stronghold.[6]

Nearly ten years later and forty-five miles to the south, the animals showed up on the radar of Morro Bay's commercial abalone divers, who occasionally ventured north to Big Sur's southern reefs. In 1948, veteran diver Lanky Tipton noticed the sea otters rafting up in a patch of bull kelp near the mouth of Mill Creek. Underwater, he found broken shells everywhere and, after several hours of searching, only a few living abalone hunkered down in deep cracks.[7] For men like Tipton whose livelihoods depended on gathering abalone, the sea otters' return was extremely bad news.

From that point on, Morro Bay divers watched in frustration as the otters moved south three to four miles each year—to Cape St. Martin in 1950, Salmon Creek in 1953, and Ragged Point in 1956—clearing each reef of abalone as they went.[8] Morro Bay abalone processor Frank Brebes reported the situation to the California Department of Fish and Game (CDFG) but got no response. State law unequivocally protected the sea otter. Eventually, the commercial divers witnessed the otters' dexterous method of hunting abalone firsthand. The lithe animals descended with a big rock in tow, used it as a hammer to break an abalone's shell, pulled the mollusk loose from the rocky bottom, and then brought it to the surface to feast.

In 1957, when the otters arrived at Beckett's Reef, the prime abalone bed north of San Simeon, the divers' frustration reached a breaking point. According to hearsay reports, as many as a hundred otters were killed, though no one was ever convicted.[9] That same year, the state's first aerial otter survey counted 635 animals between Carmel and San Simeon.[10]

Despite the reported killings, the sea otters continued to press south beyond San Simeon toward the commercial divers' most-prized home abalone beds—the nine-mile stretch of coast from Cambria south to Point Estero—an area that, at the time, produced 42 percent of all red abalones harvested in California.[11] In 1963, with increasing desperation, Morro Bay commercial divers circulated a petition requesting help from the state.

This time, the CDFG directed abalone biologist Keith Cox to conduct a one-day dive survey near San Simeon to investigate the effects of otters.[12] Until that point, most sea otter research had been conducted by zoologists watching from shore. Now, the department's divers saw with their own eyes broken abalone shells scattered on the seafloor. Cox reported it was "impossible to place a numerical figure on the amount of depredation," but "obvious that otters had been taking abalones and represented serious competition to divers on the same beds."[13]

That fall, the *Los Angeles Times* reported on the commercial divers' plight, explaining how the department's dive survey had "proved openly and publicly that the otter was a detriment to the abalone industry and could end it within the next decade."[14] However, that article was followed by a firm statement from department director Walter Shannon: the state would continue with "rigid enforcement" to protect sea otters. Shannon said that inspections by his agency indicated that no beds were being wiped out, that abalone harvests had remained stable, and that sea otters had been feeding on sea urchins, crabs, mussels, and abalone "for centuries without wiping out any of these species."[15]

The CDFG was proceeding with caution. In fact, there were considerable differences of opinion within the agency about what was happening. E. C. "Charlie" Fullerton, a former boat pilot who'd risen to the rank of inspector and would later become the agency's director, was one of the first to note the potentially positive aspects of otters' return. "In my experience, the otters' main diet is sea urchins. And this [return of otters] is the best thing that could happen. When the fishermen remove the large abalone from the rocks, the sea urchins move in and take over, leaving no room for other abalone to grow."[16]

Fullerton was expressing a newly emerging line of ecological thought. Other Sacramento bureaucrats recognized the popular sea otter as a highly political animal to be handled with kid gloves. But as the department's field biologists conducted more underwater surveys, their findings affirmed the commercial divers' contention that sea otters were indeed clearing out the state's richest remaining abalone beds.

However, a different opinion soon came to light—that overfishing by commercial divers, not sea otters, was the cause of declining abalone harvests. In a 1963 letter to the *Los Angeles Times,* John Prescott, curator of fishes at Marineland of the Pacific, a popular aquarium on the Palos Verdes Peninsula, wrote, "I think that rather than accuse the sea otter of threatening the abalone fishery, the persons involved in this fishery and those regulating it should look to over-exploitation as a cause for their problem."[17] A new debate began to quickly escalate.

Such allegations were tough for the Morro Bay divers to bear. They took their mounting frustration to state senator Vernon Sturgeon (R-Paso Robles), who helped orchestrate a special Senate Natural Resources Committee Fact Finding Hearing in San Luis Obispo in 1963. Commercial divers came out in force to explain their situation. A half dozen spoke, earnestly describing how they'd witnessed red abalone beds they'd harvested for three decades become barren after the otters moved in. The otters, they argued, didn't follow size limits or honor closed seasons but simply polished off all the abalone they found—sea urchins, too. If the otters continued to move south down the coast, they contended, the abalone industry would be lost.

Newly united as the California Abalone Association, the commercial divers and abalone processors asked for urgent state action and pressed the CDFG to translocate sea otters out of abalone beds in San Luis Obispo County. They tracked down and provided information on otter relocation work already done in Alaska in relation to the US Navy's atomic test blasts.[18] However, despite their personal accounts of the otters' underwater "devastation," the commercial divers faced an uphill battle of public opinion. Their contention that sea otters were wrecking California's abalone industry was mightily questioned.

From the outset, the Department of Fish and Game's deputy director Harry Anderson explained that his agency's position was based squarely on the research of UCLA zoologist Dr. Richard Boolootian, a scientist with impressive credentials who had meticulously observed the sea otters for ten years. One of Boolootian's key findings was that abalone made up only 5 to 10 percent of the sea otters' daily diet, with preferred urchins and mussels making up the rest.[19] By Boolootian's accounting, California's entire population of sea otters ate just 720 pounds of abalone a day—a drop in the bucket compared to the commercial divers' take of nearly eleven thousand pounds per day.[20] In addition, Anderson explained, while otters had moved into central California fishing grounds, the commercial landings of abalone had continued to increase, after a brief downturn, to more than 4.5 million pounds per year.[21] Finally, the numbers of commercial divers had also increased—eighteenfold—from 27 licensed operators in 1937 to 130 in 1947 to 294 in 1954 to more than 505 permittees statewide in 1960.[22] "When there are more fishermen, the average catch per fisherman can be expected to decline," he explained. "It is very natural for fishermen to be unhappy with a lower average catch . . . , but I do not believe we should blame the sea otter for this."[23]

However, the increased overall landings that Anderson reported were coming from the Channel Islands. Morro Bay diver Buzz Owens explained that when statewide landings peaked in 1957, over 40 percent of California's red abalone harvest had come from Beckett's Reef and San Simeon; now, just six years later, less than 3 percent came from that area. The Central Coast harvest was, indeed, plummeting.[24]

Nevertheless, a legion of scientists, residents, and conservationists spoke in defense of the sea otter, with most repeating Boolootian's finding that abalone made up only a small percentage of the otters' diet. One described massive mounds of abalone shells "dumped by processors" along the roads leaving Morro Bay as evidence of mankind's own "predator tendencies." Others described the otters' aesthetic and scientific values and made moral arguments for protecting the playful animal: it was just reclaiming its rightful place in the natural order.[25]

To the Morro Bay commercial divers who had seen the broken abalone shells at the bottom of all their best dive spots, Boolootian's figures didn't add up—and didn't really matter. From the start, abalone divers had contended they didn't want to harm sea otters; they just wanted to protect what they considered to be their own shellfish beds. They wanted a two-state solution, with a clearly demarcated boundary, preferably at the Monterey–San Luis Obispo county line.

However, at the end of the day, Senator Sturgeon conceded that it would be difficult to meet the commercial divers' demands. "In all honesty, I have no thought whatever that this Committee or the California Legislature is going to do anything with the sea otters," he explained, trying to focus the group on other options.[26]

As an upshot of the San Luis Obispo hearing, the Department of Fish and Game proposed to give the small group of Morro Bay divers some palliative relief by loosening up existing abalone harvest regulations, but these recommendations went nowhere in the legislature.[27] Commercial divers' aggravation continued to fester.

Five years later, with the state's sea otter population topping seven hundred and a herd poised to hit the Cambria–Point Estero abalone beds, the Morro Bay divers agitated once again, this time pressing state senator Donald Grunsky (R-Watsonville) to introduce legislation that would "corral" otters into the waters of Monterey County. The commercial divers reached out to sport divers, urging them to join in the legislative fight to save abalone from the sea otter, but their pleas fell mostly on deaf ears.[28] The sport divers blamed commercial divers for decimated stocks, especially in southern California, where tensions between the two groups had been rife for decades.[29]

Grunsky's senate colleagues, who felt uncomfortable with the bill that would end fifty years of special protection for the rebounding sea otter, responded with a classic legislative maneuver. They delayed action by directing the CDFG to study the feasibility of confining sea otters within their existing Big Sur refuge—north of the Monterey–San Luis Obispo county line.

The following year, the department proposed a three-year pilot study to transplant a small number of otters from the Point Estero friction zone to the northern part of their refuge.[30] At the very least, it would give some small measure of relief to the Morro Bay abalone divers.

* * *

The escalating conflict drew Big Sur artist and conservationist Margaret Owings into the debate. Owings was already a veteran of many California conservation battles. For twenty years, she had worked as a member of the Point Lobos League to protect the beaches south of Carmel River. (In 1933, after Alexander Allan died, his family sold the Point Lobos property, where he and Gennosuke Kodani had dried abalones, to create a new state park.) With her husband, noted architect Nathaniel Owings, she had campaigned to keep Big Sur undeveloped. She'd also sparred with fishermen, defeating a plan that would have dynamited rookeries of salmon-eating sea lions just north of San Francisco.[31] In addition, Owings served as the only woman on the California State Park Commission. In 1967, when she realized the state might cave on its commitment to protect sea otters, she founded Friends of the Sea Otter.

Owings was a formidable advocate and a master networker. Within its first year, the group had garnered five hundred members. She assembled an impressive board of scientific advisers from the University of California and Moss Landing Marine Labs, and she collaborated with the California Academy of Sciences in developing a sea otter photo exhibit to tour the state. She invited the famed Jacques Cousteau to film the sea otters, contacted the National Geographic Society to assist with research, and invited primatologist Jane Goodall to study the otters' distinctive use of tools. She knew just how to tap the interest of the media to spread news of the sea otters' plight.[32]

In January 1969, when a massive oil spill off Santa Barbara focused national attention on the fragility of the ocean with disturbing images of oil-soaked birds and sludge-covered beaches, Owings was struck by the extreme vulnerability of the otters. Unlike seals, which have thick layers of fat to keep them warm in frigid ocean water, otters depend on keeping their pelts impeccably clean and filled with air bubbles for warmth and buoyancy. An oil spill on the Central Coast could wipe out the small group of surviving sea otters.[33]

Recognizing the opportunity to create a second, separate population of otters to hedge against the hazards of another oil spill, Friends of the Sea Otter decided to back the Department of Fish and Game's translocation study, though the group disagreed with "corralling" otters to benefit the shellfish industry. The affluent Owings also donated funds to the University of California Santa Cruz to support additional sea otter research.[34]

Later that year, department biologists tagged and removed twenty-three otters from the Cambria–Point Estero area, building on techniques pioneered by Alaska biologists who had moved otters ahead of atomic test blasts in the Aleutian Islands. They placed seventeen animals back north up the Big Sur coast and used two to investigate the potential use of sonar barriers for containment (four others died owing to stress of handling). Within just a few months, five of the tagged otters returned to Point Estero, while still more "new" otters moved in, inexorably drawn to the abundance of abalone.[35]

With the failure of translocation, the state's options narrowed. In November 1969, Senator Grunsky hosted a conference of experts, including two sea otter biologists from Alaska. One offered his frank assessment: "To get to the meat of this thing here, you obviously have to make a decision on otter or abalone." He suggested culling the otters to maintain a smaller herd. However, even Grunsky regarded the tactic of intentionally killing sea otters as "repugnant and unacceptable."[36] Unless the group could find some course of action that everyone agreed on, he warned, there would be no action in the California legislature.[37]

Yet there seemed to be no solution. Moving otters would put them into conflict with other fisheries, and the state didn't have the financial resources to move enough otters to protect Morro Bay's abalone fleet, which had dwindled down to only twenty-five full-time divers.[38]

As the Friends of the Sea Otter argued, far greater numbers of people were now cheering the sea otter's return and willing to forgo the luxury of eating abalone steaks. The economy and culture of Monterey and Carmel—and for that matter, all of California—were shifting away from the meaty shellfish and in favor of the furry mammal.

* * *

Although commercial abalone divers saw sea otters as a threat to their livelihoods, fishery managers and marine biologists increasingly saw them as part of an ecological puzzle. If predators wiped out their prey, their own populations would crash, too—so, if abalone and otters had lived side by side on the Pacific coast for millennia, how had it worked?[39]

In 1966, Earl Ebert inherited Keith Cox's part-time position monitoring abalone for the state. Ebert had worked as a scientific diver in southern California studying artificial reefs. Though he was told by supervisors to steer clear of the contentious sea otter issue, as a scientist, he didn't see how he could avoid the matter.

Ebert wanted to get a clearer understanding of how the otters were affecting abalone, and so right off the bat he started a year-long study near Pico Creek, where a group of sea otters had just moved in. By combining observations from shore with underwater surveys, Ebert gained several key insights.

He soon realized that prior investigations, conducted with spotting scopes from shore, had not been able to consider the relative abundance of food items available to otters. Yet how the buffet was set, so to speak, made all the difference in what the animals chose to eat. For example, he never observed a sea otter eating a sea urchin, but he observed urchin "spines and test fragments profusely litter[ing] the sea floor attesting to their former presence."[40] Had his studies started right when the otters moved into the area, Ebert suspected he would have found urchins making up the bulk of the otters' diet. He also noticed the lack of other small invertebrates generally common in the nearshore marine environment—including kelp crabs, abalone jingles, and keyhole limpets—and suspected that these, too, had been eaten early on.

Reviewing all sea otter food habit studies conducted since the late 1930s, Ebert realized that the disparities reported in what otters ate could readily be explained by considering whether or not otters were feeding in fresh grounds. In long-inhabited areas, such as where Boolootian had conducted his research nearer to Monterey, sea otters consumed a wide mix of foods with lower percentages of each item. But in newly colonized zones with long-untapped foods, otters favored the most easily obtainable items that provided the greatest food value. At the San Simeon and Point Estero reefs, that meant abalone. Moreover, Ebert did his own dietary studies and found that sea otters typically ate two to four red abalones per feeding; in biomass, abalones were clearly a principal food item, making up 60 percent of the otters' diet by weight, far more than Boolootian had estimated.[41]

In October 1967, the CDFG conducted a massive underwater survey, with twenty-seven divers observing transects at ninety-one dive sites along one hundred miles of coastline from Monterey to Point Estero. The survey indicated that, after otters moved in, urchins and abalone didn't disappear but became constrained to the cracks and fissures in rock reefs. Ebert concluded that abalone populations simply could not "flourish" in the presence of sea otters.[42]

Building on early scuba survey findings by Hopkins Marine Station ecology researcher Dr. James McLean, Ebert also put his finding into a broader historical perspective. "The fact that sea otters were once plentiful suggests that in the past abalone, sea urchins, and other organisms fed upon by otters were relatively minor faunal constituents," he wrote. "It may be that many of these species sought by sea otters for food were largely confined to inaccessible crevices."[43] A natural world populated with sea otters meant fewer abalone for people.

Ebert's conclusion, reported in the *California Fish and Game* journal in 1968, represented a monumental change in understanding the abundance of abalone. It may be difficult to fully appreciate in retrospect, but it was a true paradigm shift because, for more than a century, everyone had regarded abalone *abundance* as the natural state. Ebert's findings were not recognized for their revelatory significance. Instead, they entered into a highly charged political context and were perceived to "blame" otters for eating abalone.

Many years later, Ebert would recall, "At the time, everyone saw the sea otter as a beautiful animal, and no one wanted to hear that its eating habits were having an impact on shellfish, including the higher-ups within the department and, of course, Friends of the Sea Otter."[44] When he'd started work at the department in the late 1950s, Ebert distinctly recalled there had been an expectation that fishery management would follow from research conducted by CDFG biologists; but with all the controversy and rancor over abalone and otters, the science itself became contested and political. Ebert recalled that people at public meetings would stand up, shout, and accuse him of not knowing what he was talking about, even though he was the only one who had conducted underwater studies focused on the issue. "People simply did not believe the otter was having an impact."[45]

The department had tried to keep its sea otter studies separate from its abalone monitoring, and despite Ebert's research that showed the two animals were inextricably linked in a predator-prey relationship, the intense social politics put the agency on the defensive. Ebert's studies illuminated emerging ecological perspectives yet were also used to justify the fishery management perspective held by many in the department who wanted to cull otters to balance the clashing demands of different groups. Familiar philosophical moorings were adrift.

For these reasons, the new ecological way of thinking about abalone and otters did not compel the CDFG to reexamine the assumptions on which the abalone fishery had been—and still was—managed. Few seemed to recognize that the department's purported success at managing abalone for a sustained, high-yield harvest for nearly a century with size limits had also been the artifact of a mistaken baseline, a windfall of "virgin stocks," and a series of external social conflicts that had afforded the mollusks timely periods of reprieve to grow despite ever-expanding fishing pressure. Rather, the department retrenched and aimed primarily to avoid conflict.

* * *

Commercial divers blamed the decline of central California's red abalone squarely on sea otters, but there was no such clear culprit behind the decline of abalone on southern California's mainland coast.

In the 1950s, Dr. Wheeler North, a young research biologist at the Scripps Institution of Oceanography in La Jolla, had been hired with a grant from the Department of Fish and Game to study southern California's giant kelp— stands of *Macrocystis pyrifera* that grew in massive underwater forests. North was among the first of a new generation of scientific divers. He had bought one of the very first Aqua-Lungs sold in the United States and wore the new-fangled scuba gear with woolen underwear before wetsuits even existed. North's job was to inventory the kelp forests, study their biology, and assess the effects of the kelp industry, which harvested fronds for algin—used to make food products from beer to cosmetics to ice cream. For decades, fishermen had contended that industrial kelp harvest was damaging southern California's coastal fisheries, including abalone. North began by first documenting the historic extent of the kelp beds with old maps and aerial photos, and then he started some of the earliest underwater kelp forest surveys at Point Loma off San Diego. Though he would ultimately conclude that the effects of kelp harvest on fisheries were negligible, North found that storms could have significant impacts.

It just so happened that, during the course of his study, the powerful 1957–1958 El Niño brought its flood of warm waters and potent storms into southern California. El Niño was still unknown as a recurring oceanographic phenomenon; rather the warm influx and storms were considered to be damaging aberrations. Elevated water temperatures persisted for three years—the most significant sea surface temperature change on record at the time—resulting in a massive kelp die-off. North determined that the nutrient-poor warm water weakened the kelps, making them vulnerable to damage by storms that forcefully tore the plants from their holdfasts.[46]

North's pioneering scuba research soon expanded to include underwater investigation of southern California's sewer outfalls that funneled increasing amounts of effluent and industrial pollution from booming cities into the marine environment. Near White's Point outfall, which flushed Los Angeles' wastewater to sea just south of the Palos Verdes Peninsula, fishermen had reported kelp forests dying. Underwater, North and his colleagues found vast expanses of seafloor devoid of kelp and completely blanketed by red and purple sea urchins. At White's and at other outfalls, they found "urchin barrens" so dense they contained fifty soft-ball-sized red urchins per square meter.[47]

In the giant kelp forests, *Macrocystis* converted sunlight into a primary food source. Creatures such as urchins and abalone ate the rich alga. Then, fishes like sheephead ate these herbivores, and so on, up the food web to larger fishes, sea lions, and, ultimately, sharks. The emerging science of ecology was revealing kelp forests to be the foundation for a vibrant and valuable community of marine life.

North and his colleague Dr. John Pearse wondered how the urchins persisted in such high densities even after the kelp was grazed down and completely eliminated. They ultimately determined that amino acids from the sewage dissolved in seawater and served to nourish and sustain the large urchin populations. This meant the kelp never had any chance to recover. In their 1970 paper, published in *Science*, they also recognized extirpation of the sea otter (a primary urchin predator) and depletion of abalone (a primary urchin competitor) as two factors that had also likely played into the urchins' dominance, causing troubling but yet-to-be-understood disruption of southern California's underwater ecosystems.[48]

The research of McLean, Ebert, North, Pearse, and others was beginning to reveal that marine ecosystems were affected by a multitude of factors—elevated ocean temperatures, storms, and pollution, but also biotic interactions, such as competition and predation, within the community of creatures.

The powerful influence of these undersea community interactions was further illuminated by research in Alaska, where the Department of Defense had dispatched biologists to analyze the impact of atomic blasts on the marine environment. While the sea otter–abalone controversy was reaching a boiling point in California, Alaska provided a natural laboratory free of such social complications. Some islands had otters. Some did not. Around islands with no otters, the seafloor was carpeted with sea urchins; around islands with otters, lush groves of kelp buffered waves and harbored a plethora of invertebrates and fishes. The biologists found the differences to be breathtaking.

In the mid-1970s, based on this field research, researchers James Estes and John Palmisano would articulate the hypothesis that sea otters, as top predators, played a significant role in structuring the composition of the entire nearshore marine environment.[49] Otters exerted "top-down" control of urchin populations and thereby reduced grazing pressure, which enabled kelp forests to grow luxuriantly and to provide food and habitat for a rich array of marine life. The sea otter also exemplified what ecologist Robert Paine had recently described as "keystone species"—organisms that played a pivotal part in governing the stability of a whole ecosystem.[50]

Estes and Palmisano's work in Alaska, also published in *Science*, suggested that the return of sea otters to California's coast might help restore degraded kelp forests, even if it meant urchins and abalone would be consigned to dwell within protected crevices. They concluded that the otter was "an evolutionary component essential to the integrity and stability of the ecosystem."[51] Later researchers would contend that southern California's kelp forests have greater complexity, with far more predators and prey species than in Alaska, and cautioned against drawing conclusions based on distant ecosystems. Yet the general principle that top predators could influence ecosystems—and that sea otters could play a crucial role—was gaining credence.[52]

These new ecological findings were coming to light at the very time that California's fishery managers, lawmakers, and citizens were struggling to understand and address conflicts heightened by ever-increasing fishing pressures, the degradation of marine ecosystems in southern California, and the return of sea otters in central California. Yet the science remained inchoate and was not yet sufficient to inform policies or quell hostilities brewing in social and political realms.

Up until that point, fisheries managers had been trained in the science of maximizing sustained commercial landings based on the simplistic assumption that fish—and shellfish—reproduced and grew at regular rates and could yield a harvest each year like corn, soybeans, or trees. But now, the science of ecology was revealing far greater complexity. Fluctuating temperatures, plus pollution, fishing pressures, and the ever-changing structure of biotic communities were not just backdrop, but active, interrelated forces affecting marine fisheries.

* * *

In the spring of 1970, with sea otters starting to consume abalone at Point Estero's valuable reefs, the exasperated Morro Bay commercial divers persuaded Senator Grunsky once again to introduce a bill to corral the sea otters and to allow for occasional "take." This time the Department of Fish and Game favored the commercial industry's plan.[53]

Although the CDFG's leadership had initially taken a strong stand to protect the sea otter, additional field studies, growing numbers of otters, falling abalone landings, and shifting state politics changed its official bent. Over time, the department's fisheries biologists had come to regard the predators as the prime culprit in the decline of abalone landings. The drop was sharpest at Morro Bay, where the abalone harvest had fallen from 1.4 million pounds in 1967 to a mere 414,000 pounds in 1969.[54] "The way the data looks to me," department biologist Jim Messersmith explained, "the sea otter has expanded and the abalone catch has declined rapidly."[55] Despite the failed translocation experiment, department biologists still believed that a well-crafted plan to capture and move otters to the coast of Washington could accomplish the aims of conserving both otter and abalone.[56]

However, Friends of the Sea Otter regarded the CDFG's emerging strategy of "active management" as a euphemism for killing otters that were still vulnerable. They considered the thousand remaining animals to be a small vestige of a much larger historic population and remained convinced that commercial divers—not sea otters—were responsible for declines in abalones.[57] Margaret Owings was adamant: "The truth is that they have depleted this resource by over-fishing."[58]

In the weeks before the California Senate Natural Resource Committee hearing, sea otter advocates prepared for a showdown. Owings organized and sent a mail alert to sixteen thousand supporters, urging them to write and tell their state senators to oppose Grunsky's bill.[59]

In early April 1970, just two weeks before campuses and communities all across America celebrated the first Earth Day, Morro Bay abalone divers and sea otter supporters crowded into the state capitol for what would be the final hearing. Commercial divers brought a hand-drawn map showing the march of sea otters down California's coast, and diver Ernie Porter pleaded with the senate committee: "We're going down the drain and we know it. . . . We're dead ducks."[60] Gloria Pierce, a Pierce-family cousin representing the Morro Bay abalone industry, explained, "We are not asking that the sea otter be destroyed, we are instead asking that it be managed for its own protection as well as the preservation of our shell fish resources."[61]

The sea otters' formidable cast of "friends" also spoke, representing the changing tenor of the times. William Penn Mott Jr. of the State Park and Recreation Commission explained that sea otters were a favorite with tens of thousands of visitors to Point Lobos Reserve and "must be protected from decrease." Dr. Robert Orr of the California Academy of Sciences charged that the misguided measure was proposed solely so "fishermen can continue their overexploitation to produce a gourmet food item."[62]

When the contentious hearing ended in a standoff, Senator Robert Lagomarsino (R-Ojai) recommended postponement so that a compromise could be reached "that would satisfy both sides."[63] Three weeks later, Senator Grunsky withdrew his bill, recommending again that the committee pursue further study. Margaret Owings regarded the tabling of the bill as a victory.[64]

With the failure of the Grunsky bill, the Department of Fish and Game lost its appetite for dealing with sea otters. Later that year, Charlie Fullerton, the department's new deputy director, reported that his agency would make no further recommendations regarding otters until the three-year study started in 1968 was completed. "The more we learn about the otter, the more we realize that we know very little," he explained. "We want to retrench and see if what we are doing so far is right."[65]

Delays would make the contentious issue moot. Sea otters continued to move south, eating exposed abalone. By 1971, commercial divers were compelled to abandon diving grounds north of Morro Bay.[66]

* * *

In the summer of 1971, reporter Jane Eshleman Conant broke a story in the *San Francisco Sunday Examiner and Chronicle* about the Department of Fish and Game's long-awaited sea otter study. She predicted that it would make "bombshells go off in all directions."[67] I found the clipping of Conant's article in an archived file of then department director Ray Arnett, indicating the agency's heightened sensitivity about the otter-abalone conflict. The study's official release would be delayed, but Conant's description of what it promised to divulge is revealing.

The "theory is this," she explained: in the past, "with an estimated 200,000 of the affectionate and playful otters living along the shores around the entire Pacific Basin, from the top of Baja California, to Japan via Alaska, their prey huddled in crevices in rocks." During this prehistoric period, she continued, the otters couldn't reach most abalone and probably ate other foods like urchins and fish, but then, after the sea otter was hunted to near extirpation, the abalone population had "exploded."

In the vacuum of the department's politically imposed silence, Conant explained a critical irony to the broader public: that there would have been no commercial abalone fishery—no juicy abalone burgers or savory abalone sandwiches—if the sea otter hadn't been wiped out in the first place.

With the population of otters in California now reaching 1,200 animals, Conant continued, "the abalone and crabs are probably retreating into crevices again, reestablishing the balance of 2 centuries ago. But now this ancient balance is upset by the demands of a huge and hungry human population. The question: who has priority—the otters or ourselves?"[68]

Conant was among the first to articulate in the popular media the new ecological way of thinking about abalone and sea otters—and also the more profound questions it raised about the place of humans in the natural world.

The return of the sea otters finally revealed that what many Californians had considered their birthright—the opportunity to hunt and eat from an abundant natural supply of abalone—had, in fact, been an artificial and temporary peak in abundance, created by a series of historic events that occurred long before anyone could begin to piece together cause and effect.

✳ ✳ ✳

In the years that followed, public support for sea otters continued to grow, especially in the wake of atomic tests in Alaska's Aleutian Islands. In November 1971 one such test, the massive Cannikin explosion—385 times more powerful than the bomb dropped on Hiroshima—blasted Amchitka Island, a national wildlife refuge established in 1913, in part to protect the sea otter. Initially, Atomic Energy Commission officials reported that only eighty sea otters had been killed; later the toll was upped to one thousand.[69] That same fall, the popular TV series *The Undersea World of Jacques Cousteau* aired, as its season's premiere show, "The Unsinkable Sea Otter," featuring footage of adorable sea otters from Monterey.[70]

Meanwhile, the widely publicized drowning of dolphins in tuna fishermen's nets added to a growing sense of outrage that marine mammals were under siege, spurring Congress to pass the Marine Mammal Protection Act in 1972. The new law officially put California's sea otters under the jurisdiction of the US Fish and Wildlife Service, thereby overriding the management plan that state wildlife biologists had continued to develop to contain sea otters in Monterey County waters. In 1977, the sea otter would be listed as federally threatened species under the Endangered Species Act.[71]

The new federal laws put an end to the Central Coast's abalone–sea otter conflict in a legal sense, but the anguish and animosities felt by commercial divers, who believed the shellfish belonged to them, would persist and morph into a deep distrust of the state agency that had previously served their interests. Moreover, the prospect of sea otters expanding into southern California's waters would continue to loom over all decisions regarding management and fishing of abalone.

Despite abalone's long-hallowed place in California cuisine, most Californians would relinquish the shellfish with little fanfare—preferring to be the people who saved the sea otter rather than who savored a seafood.

The return of the sea otter had provided a key to unlock a new ecological way of thinking about abalone abundance and would wholly reshape central California's nearshore marine environment. But otters were not the only reason for the abalone fishery's decline, which was now poised to accelerate from steep to precipitous.

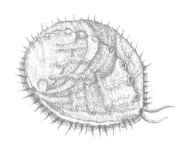

CHAPTER 8
Last Best Chance

Growing up in the late 1950s in southern California, Gary Wickham and his childhood friends trekked each day to La Jolla's Bird Rock tide pools, where they'd pry an abalone off the eroded sandstone reef, slit its foot, and then watch as all sorts of fish came streaming in to chow. "It was probably illegal," he now concedes, but as a boy, he felt like he was at the center of a nature movie. The kids brought guidebooks down to learn the tide pools' plants and animals. Soon they began foraging.

Each week, Gary brought abalone home for his mother to cook, a mound of shells growing in their backyard. While kids elsewhere were packing peanut butter and jelly for school lunch, Gary and his friends' sandwiches were filled with abalone.

By the time he came of age, Gary was fully immersed in southern California's beach culture, hanging around with a scantily clad group of tanned and towheaded surfer kids from the neighborhood. Rejecting societal expectations, these early counterculture youth partied hard and embraced a concept of vigor that entailed regular immersion in the ocean, which they regarded as the "real world" and reverently respected for its awesome power.[1]

Abalone were central to the La Jolla scene—a free food that perfectly complemented the freedoms afforded by warm water and sunshine. Its meat was rich, nourishing, easy to cook on the beach or in someone's backyard, and the whole experience of foraging and pounding felt primal. By the late 1960s, the shells' iridescent swirls enhanced psychedelic trips. Abalone could even be sold if someone needed a little extra cash. It wasn't legal, but at the time, all rules seemed to Gary and his buddies like suspect constructs of an authoritarian adult world.

By the 1970s, with far fewer abalone on the nearshore reef at Bird Rock, the La Jolla gang, now calling itself the Mac Meda Destruction Company, would pack a boat with fifty friends and a keg of beer and head out to free dive on deeper reefs. They'd skin dive to take their daily limits of five each—bringing a boatload of abalone back for raucous backyard parties, echoing earlier iterations of beach culture from the Carmel bohemians to the Bottom

Scratchers. Gathering and eating abalone meant friendship, freedom, joy, and connection to the ocean.[2]

Abalone had come to carry similar if more genteel meaning in middle-class culture, too. Through the 1950s and 1960s, *Sunset Magazine* published many articles advising the rising tide of newcomer Californians how to brave "chilling surf and surging offshore currents—in search of these tasty, rock-clinging shellfish," with recipes such as Barbecued Abalone, Abalone and Crab Hollandaise, and Abalone Pot Roast in Wine.[3]

But in 1972, an article about how to make "Abalone Beach Chowder" would be the last. It recommended taking along two to three cans of chopped clams as a safeguard "if hunters come back empty handed."[4]

At some point, Gary remembers he and his friends just stopped taking abalone because they saw too few left and knew the fishing had become "excessive." As a young man, he'd had a recurrent dream of diving for abalone. In his subconscious state, he'd sense the push and tug of pulsing waves and see in his mind's eye the studded pattern of oval shells on rock. Then the abalone disappeared. He never had the dream again.

* * *

During the years that the Department of Fish and Game was enmeshed in the sea otter conflict in central California, abalone fishing expanded in southern California. Sport divers continued to hunt with scuba gear, and with former abalone beds turned to sea otter habitat in central California, commercial divers were compelled to focus their fishing on submerged reefs surrounding the Channel Islands. In the late 1960s, the advent of powerful Radon boats (named after commercial-diver turned boat-builder Ron Radon) made the islands' once-remote abalone stocks far more accessible; plus, the faster and lighter boats could carry heavier loads back to market, allowing divers to stay out working for longer periods.[5]

Since 1916, California's Department of Fish and Game (CDFG) had been tracking abalone with commercial landing data—the poundage that fishermen reported harvesting when they returned to the dock. Traditionally, fisheries managers had presumed that as long as fishermen continued to haul out tons of fish and shellfish, there were still plenty more left to take. In the context of abundance, it was a shorthand way to know the status of stocks without the time-consuming and expensive research required to know each animal's biology. But in retrospect, it was like balancing a checking account by looking only at what you are taking out.

With heavy fishing through the 1950s and 1960s, overall abalone landings had remained high, bolstered as commercial fishermen tapped red and then also pink abalone beds on the Channel Islands.[6] But by 1969, as a larger fleet of dive boats fished the relatively small area, stocks of those abalone wore thin and landings started to drop—precipitously. In response, the state Fish and Game Commission finally, at the request of the abalone industry, instituted a $100 fee for commercial permits to discourage more divers from joining the fishery. As a protective measure, the commission also increased the size limits for green, pink, and white abalone, each by a quarter inch, to buffer against overfishing.

However, the commission's primary response would be to protect the fishery rather than the abalone. To forestall the decline in overall landings, commissioners opened shallow areas around the most remote islands to give fishermen access to untapped beds; then they decreased size limits for the smaller and less-common pinto and flat abalone, and reversed course to decrease size limits for greens. The tactic of adjusting size limits by quarter-inch increments was clearly inadequate, but at the time, there were few "tools" in the abalone management toolbox. Responding to the rule changes, the fishery shifted to target different species until stocks of all were reduced.[7]

Most consequential, in 1971, with black abalone in the intertidal zone as the only abalone species left in large numbers, the commission rescinded the abalone export restriction that had been in place since 1913 to allow commercial fishermen to take these smaller, tougher animals for the Asian market, which still favored dried abalone. Before long, most of the commercial catch of abalone in California would shift to supply markets in Japan and China, where unique and still enduring cultural values meant much higher prices.[8]

By 1973, California's commercial abalone harvest had dropped to 1.3 million pounds—less than one-third the amount landed ten years earlier.[9] If landings numbers were the pulse that showed the promise of a fishery, the once steady beat was now weak.

Finally, the state Fish and Game Commission directed the Department of Fish and Game to revive its research on abalone. The department hired biologist Richard Burge as project leader, directing him to figure out why southern California's fishery had crashed and to develop a plan to restore stocks and increase yields within two years.[10] A young man with a crest of reddish hair and a trim mustache, Burge took on the project with gusto. After studying shellfish biology at Humboldt State University, he'd already garnered diving experience by working with department biologist Earl Ebert on abalone monitoring.

From the get-go, Burge recognized that abalone species in each area of the state faced different environmental constraints and pressures, yet the state's

8.1. After the abalone fishery collapsed, the California Department of Fish and Game hired Richard Burge to head up a research and recovery project in 1973. Burge identified overfishing, including damage from bar cuts, as a prime reason for decline and recommended ratcheting back fishing pressure, but the legislature took negligible action. (Photo courtesy Richard Burge)

one-size-fits-all management approach did not reflect those nuances. Despite earlier research by Bonnot in the 1940s and Cox in the 1950s that identified highly variable growth rates, no one had yet managed to collect the growth data needed for refined management of southern California's abalones. To properly manage the wild shellfish, Burge knew he'd need an accurate inventory of specific abalone populations, growth information by species and location, more knowledge about each species' life history, and, most important, a better understanding of productivity—how many abalone actually "recruited" into the population of up-and-coming animals each year. He envisioned the ambitious goal of modeling the abalone fishery, which would enable the department to carry out more responsive, place-based management for the first time.[11]

Starting in 1974, Burge led sixteen research cruises over an eighteen-month period. Every two weeks, a crew of four young scientific divers headed out to the Channel Islands on the CDFG's research vessel, the *Kelp Bass*, for an eight-day stint. The divers counted abalone and evaluated their age classes; cohorts of small abalone indicated successful reproduction in the preceding years. They also tagged abalone at each site to determine how growth rates varied by location, age, and size. The first season, they tagged more than seven thousand animals, intending to retrieve them the following year.[12]

Long before they could re-collect the tagged animals, they stumbled on an important insight. To tag abalone, a diver had to quickly and carefully slide his pry bar under each mollusk, pop it off, place it in a mesh bag, and then pass the full bag up to the surface. On the boat's deck, crewmembers recorded species, lengths, and weights and threaded a stainless-steel wire ID tag through respiratory pores in the animals' shells. Then the diver resubmerged and placed each tagged abalone back on suitable rocky habitat as near as possible to where it had been collected.

There came the surprise. While replacing abalones, crew diver Mel Odemar noticed a giant bat ray perched on a ledge right above him, staring down, watching his every move, waiting. As soon as he moved aside, the bat ray swooped down and swallowed most all the abalones whole before they had a chance to reattach and take shelter in their shells. On another dive, Burge was accosted by an aggressive bat ray that snatched a full bag of tagged abalones. He grabbed the bag back, but the ray used the strength of its muscular, four-foot-wide body to wrest it away once again. The divers also confronted frenzied sheephead—large black-and-red toothed fishes that crushed and ate whole abalone.[13]

Burge figured the scent of abalone blood attracted the predatory rays and fish. Despite taking great care with their pry bars, the biologists found that it was difficult to avoid cutting an abalone's foot. And even small cuts would bleed continually because the abalone's blood did not clot.

During earlier dives on the North Coast, they'd noticed lots of red abalone—just short of legal size—lying helpless on the seafloor after a weekend of heavy sport diving. This was highly unusual behavior for abalone, which typically right themselves and immediately clamp down hard. Back in the lab, on closer observation, they'd found that the helpless animals had tiny cuts on their feet, presumably made by unwitting divers' pry bars. Burge followed up with an experiment, intentionally making half-inch cuts in a set of abalone. Roughly 60 percent died, and those that survived did not grow or spawn for at least ten months.[14] Between bleeding and predation, tiny cuts were a big problem.

The high mortality caused by picking and replacing abs had not been noted by previous researchers, but the Burge crew experienced it every time they tried to tag abalone in southern California. When they returned to check a site that they'd worked the previous day, they found stainless steel tags sitting in piles of masticated shell debris. Burge's crew determined that a significant portion of the abalone they removed and tagged died as a result of being picked, returned, and then snapped up by predators. In the case of more delicate pink abalone, the mortality rate was as high as 30 percent.[15] They changed their protocol—they stopped replacing inadvertently cut animals and assigned one diver to chase off aggressive rays with a bang stick.

More important, Burge and his crew recognized the broader ramifications of their findings. To meet the state's size-based fishing regulations that required divers to return undersized abalone, thousands of divers were constantly picking and replacing "shorts." If trained and experienced CDFG divers were causing mortality up to 30 percent, thousands of novice sport divers and new commercial divers could be causing mortality up to 90 percent.[16] Burge realized that far more abalone were being killed in the fishery than the department had accounted for in commercial landings.

By the end of 1974, the biologists drafted their plan for the Fish and Game Commission. Though never officially published, a mimeographed draft of what became known as the "Burge Report" became the department's unofficial synopsis of the abalone fishery's precipitous demise and set the framework for public discussion about what could be done.[17]

The biologist team identified "excessive fishing pressure" as the foremost cause of decline. Pry-bar cutting by too many divers—both sport and commercial—plus escalating efficiency of gear had led to overfishing of an ever-shrinking pool of abalones. Between 1965 and 1973, there had been a 400 percent increase in recreational party-boat divers in southern California and a 250 percent increase in their catch, as well as greater take by rising numbers of central and northern California sport divers, too.[18] In the commercial realm, during the 1950s, there had been about seventy-five abalone fishing boats, but by the early 1970s, there were more than two hundred, with 360 commercial divers. Moreover, each year up to 50 percent of commercial divers were entirely new to the fishery—mostly college students and sport divers picking up ab diving as lucrative part-time work. "Illegal activities" by some few who took large numbers of undersized abalone added to the onslaught of fishing pressure most everywhere abalone lived.[19]

As the Burge Report explained, the decline in landings owed in large part to a downward adjustment in harvest rates that reflected a significant shift in conditions on the reefs around the Channel Islands. When the commercial fishery had opened in the 1940s, divers found large unfished populations everywhere there were reefs and kelp—the result of decades of reproduction, buildup, and gradual expansion of abalone from cracks and crevices out into open areas in the absence of otter predation. After the initial heavy take of the largest abalone and continuing fishing pressure on up-and-coming animals, spawning stock had been markedly reduced. As a result, in areas hardest hit by the fishery, the Burge team had found poor larval recruitment and some areas completely denuded. Beyond reduced spawning, Burge speculated that, with fewer adult abalone, fundamental changes to community structure had occurred in some places. As urchin barrens and encrusting fauna took over rock surfaces, less protected nursery and rearing habitat remained for juvenile abalone.[20]

Whatever the case, in many areas, recruitment was now insufficient "to sustain a continuous high harvest."[21] The high landings of the inherited windfall fishery were a thing of the past, and the fishery would have to adjust to a much lower rate of take to sustain itself in the long term.[22] The Burge team's report articulated for the first time how the new understanding of past abalone abundance would require new expectations and big changes in fishery management.

Beyond fishing pressure, the Burge team counted sea otter predation on the Central Coast as the major factor that had markedly reduced commercial landings but noted that federal jurisdiction precluded any state management. The biologists also identified pollution and "urchin encroachment" as factors that affected larval development and reduced nursery habitat for juvenile abalone on southern California's mainland coast. These issues were beyond the scope of the Burge Report, but in several locales, abalone had been entirely lost, leaving no animals to parent a next generation.[23]

To address this confounding suite of problems, the Burge team recommended an "integrated" package of measures to reduce fishing pressure and to increase the abalone supply. A key recommendation was a "limited entry" system—cutting the total number of commercial divers down to seventy—more in line with the fleet size of the 1950s, when harvests by traditional "hard-hat" divers had seemed sustainable. At the time, limited entry was a new fisheries-management approach. Burge and his colleagues thought it could save the livelihoods of experienced, full-time commercial divers while pruning out the part-timers who presumably posed the highest risk for cutting abalone with pry bars.[24]

For both the commercial- and sport-fisheries, the Burge team recommended reducing the season from ten to six months and splitting the season to keep divers out of the water during winter, when rough conditions made cutting with pry bars most likely.[25]

To reduce fishing pressure in the sportfishery, the Burge Report proposed lowering daily bag limits north of Point Conception from five abalones to four. Although the sportfishery was smaller in overall magnitude than the commercial fishery and difficult to track, increasing numbers of recreational scuba divers were exerting considerable fishing pressure in accessible areas.[26]

To increase the abalone supply, the Burge team recommended shrinking size limits, a measure predicted to provide a two- to fifteenfold increase in the number of abalone available to divers. Through their studies, they'd found that the growth rates of pink, white, and green abalone seemed to drop off once the animals reached a certain threshold: it could take years for a green abalone to grow the final one-quarter inch to reach legal harvest size. The biologists speculated that the current size limits might actually be encouraging the slowest-growing abalone to persist as broodstock while fisheries culled out the fastest-growing animals. Meanwhile, the "short" shellfish took up space on the reef at their just-under-legal size, becoming especially vulnerable to bar cuts by novice divers. Lower size limits would take best advantage of new information about growth rates and still give animals six to eight years to spawn before reaching the new legal size.[27]

However, Burge and his colleagues warned that lower size limits could also lead to a "short-lived 'bonanza' fishery" were it not paired with their most important recommendation: to put a ceiling on the total commercial catch.[28] Firm landing quotas were critical to boosting the base population of breeding abalone parents. The Burge team recommended reducing commercial take by half in the first year, believing it would lead to higher harvests in the future. Landing quotas would thereafter be set each year after a review of the previous year's landings, catch effort, and field surveys. In addition, the biologists suggested tracking abalone landings in numbers of animals rather than in pounds.[29] The goal was to be more responsive to changes in actual conditions of the abalone beds.

The Burge team also recommended a new program to rebuild stocks, especially off southern California's mainland, where the combined effects of environmental degradation, urchin encroachment, and heavy fishing pressure had left many reefs completely barren. The CDFG had already started a pilot project to mass-produce seed abalone at its Granite Canyon laboratory south of Carmel. A new abalone mariculture program could enable the department to refine techniques, to increase production of seed abalones for restocking, to create a demonstration ocean-bottom abalone farm to serve as a model for commercial enterprise, and to investigate the possibility of leasing offshore areas for commercial abalone cultivation.[30] If mariculture worked, abundance might once again be restored—at least outside the sea otters' range.

The Burge team's plan aimed to accomplish the seemingly irreconcilable tasks of bringing abalone back to southern California while also allowing for continued exploitation. Since Charles Edwards first proposed abalone refuges back in 1913, and Bonnot recommended rotating commercial harvests in 1941, this would be the third time biologists from the state's fishery agency put forth a plan to better manage California's abalone.

CDFG acting director Charles Fullerton presented the Burge team proposal to the Fish and Game Commission in January 1975. Then Burge and his colleagues held public meetings in Santa Rosa, Santa Barbara, and San Diego to garner input. The crowd in Santa Rosa, dominated by sport divers, was disgruntled about reduced bag limits and the shorter season.[31] Many regarded poaching—not even mentioned in the Burge Report—as the most pressing threat to abalone in their region and wanted bigger fines and more enforcement officers (at the time, the department did not regard poaching as having a "great impact" on overall abalone numbers).[32] Ultimately, sport divers begrudgingly agreed to the new reduced bag limit "if the Department deemed it necessary," Burge reported.[33]

At the Santa Barbara and San Diego hearings, commercial divers and processors dominated. They supported limited entry but strongly objected to smaller size limits, reduced seasons, and landings quotas, contending that more research and a more accurate system for tracking landings were needed first. The California Abalone Association (CAA) offered its own recommendations: new log books to provide better data, more stringent diver training, elimination of the twenty-foot rule that prohibited commercial diving in shallow areas, and opening a reach of coast north of Santa Cruz to spread out commercial fishing pressure. They supported mariculture to replenish depleted broodstock but not the idea of leasing offshore reefs to create private abalone farms. The industry also argued against the smaller size limits because commercial divers believed in the traditional approach and processors found it more economic to process larger animals.[34] Finally, with sea otters continuing to move south down the coast, the CAA again proposed corralling the animals—a highly controversial measure that had been considered and roundly rejected before.[35]

Burge and his team had emphasized that their suite of recommendations was designed to work together as a whole and not to be cherry-picked. "Their effectiveness requires that the concepts of the entire program be adopted and not accepted or rejected on an individual basis," the report admonished.[36]

The Fish and Game Commission followed suit, readily adopting the Burge team's recommendations for the sportfishery. Commissioners cut bag limits and adopted new rules to reduce bar cutting—requiring divers to keep the first abalones picked rather than "shopping" for larger ones and to use proper abalone irons with smooth, rounded edges, rather than the commonly used assemblage of knives, screwdrivers, ground-down auto spring leaves, and even a popularly sold "all-purpose dive tool" with a damaging curved bar "almost guaranteed" to cut animals.[37] The commission also closed southern California's mainland coast—from Palos Verdes Point to Dana Point—to all abalone diving, with the goal of restoring the area. Since it was already "nearly depleted of abalone stocks," Burge explained, no one objected to the closure.[38]

However, the legislature, which still held purview over the commercial fishery, did not heed the Burge team's warning. As details were hashed out in the political realm, lawmakers were reluctant to adopt any provisions opposed by industry. Because the abalone industry strenuously disagreed with many of the game-changing proposals the biologist team recommended, the Department of Fish and Game political staff believed that it needed to compromise to make any headway. Everyone at the negotiating table agreed to limited entry and increased license fees to fund outplanting of mariculture-raised juveniles, but all other measures would have to wait.[39] Even the limited entry system was to be accomplished only "by attrition," with a moratorium on new permits; moreover, the recommended target of seventy divers was raised to two hundred. After meeting with CAA representatives, the department dropped recommendations for reduced size limits, shorter seasons, and landing quotas, and then arranged for introduction of a modest, limited-entry-only bill in the legislature. The toothless bill passed the assembly and senate by unanimous votes early in 1976.[40]

While sport divers pointed fingers at commercial divers, and commercial divers pointed fingers at sea otters, and everyone lamented pollution, at least one critic would blame the Department of Fish and Game's subjugation "to political and economic pressures that work against its taking a hard line on the excesses of the commercial and SCUBA sport divers" as the prime factor in the decline of southern California's abalone.[41] Despite department biologists' efforts to use science to guide management, the system for conserving the public's "fish and game" was so politically girded that it was fundamentally rigged to fail.

Burge was compelled to regard the new abalone law as a first step, and he continued to harbor hopes for a science-based approach that could rebuild existing abalone populations in southern California, model growth and recruitment, and create a responsive, long-term harvest plan based on annual quotas for each species.

However, in the mid-1970s, California hit hard economic times. The passage of Proposition 13 in 1978 limited state property-tax revenues and reduced departmental operating budgets while other legislation created specific mandates for how fish and game license revenues were to be spent.[42] Funding for science research and monitoring was heavily curtailed. Moreover, new leadership at the department made a strategic decision to refocus the state's limited abalone funds toward outplanting of farm-raised juveniles, effectively abandoning the department's wild abalone research and management program. Burge was reassigned to supervise the Morro Bay Fish and Game District Office. As he would later recall, the state abalone program he'd led was, in effect, "cut off at the knees."[43]

The CDFG's new direction on abalone likely rested on the belief that sea otters would continue to expand their range and ultimately make management efforts pointless. In the meantime, outplanting juveniles had promise as an interim solution.[44] Burge further reflected, "Commercial divers expected to get their way, and policy people in Sacramento were pretty well convinced that they couldn't do anything. And they didn't believe that the resources were really in trouble. They basically said, 'Let's just keep doing what we are doing until there is a crisis.'"[45]

Burge tried to soldier on in his abalone research, but ultimately the demands of his supervisory job left him unable to accomplish much. By 1980, he realized that his vision of using science to better manage the wild abalone fishery in California would go unrealized. "When they decided to stop research and just plant out abalone—something that just 'looked good,'" he later told me, "the writing was on the wall."[46] Burge found a new job managing wild shellfish for the State of Washington and never looked back.

After California's 1976 abalone law was enacted, UC Santa Barbara social scientists conducted a survey and found that 58 percent of sport divers and 65 percent of commercial divers described the Department of Fish and Game's management of abalone as "poor" or "very poor."[47] Ironically, divers held the CDFG fully accountable for the disappointing decline of abalone, but legislative oversight of commercial fisheries meant the department, in fact, had no authority to take the protective actions its biologists deemed necessary.

Given the commercial divers' dim view of the department, the social scientists also analyzed how responsive the agency had been to their suggestions. They found that the majority of the abalone industry's proposals since 1969 had, in fact, been adopted: size limits were reduced on green abalone and increased on pinks; the shallow-water fishing prohibition was suspended in parts of the Channel Islands; closed seasons were adjusted to their preference; the long-standing prohibition on drying, canning, and exporting abalone had been repealed to allow for more take of intertidal black abalone; research to determine the feasibility of planting out seed abalones for restoration had begun; and all objectionable portions of the Burge Report's recommendations—shorter seasons, reduced size limits, a smaller fleet, and landing quotas—were dropped. Nevertheless, the CDFG's resistance to opening the North Coast to commercial harvest and its inability to resolve the intractable sea otter issue led to "the commercial diver's belief that the department is unresponsive to his problems."[48]

With their research, the Burge team had put together more critical pieces of the abalone puzzle. Their study and recommendations would be the department's last attempt to stave off the decline of southern California's abalone fishery beyond the range of sea otters, but the watered-down legislation ultimately enacted was already too little, too late.

There was no way to go back to the heyday of abalone superabundance, but political intransigence meant a crucial opportunity to proactively manage the remaining windfall fishery and to conserve California's abalone into the future had been squandered.

CHAPTER 9
A Wishful Tack

Ever since Charles Lincoln Edwards had sounded the alarm about abalone "extinction" in southern California and first tinkered with placing black abalone in an early preserve off the Venice Breakwater in 1910, scientists, divers, and citizens had periodically called for the State of California to try mariculture as a way to rebuild depleted abalone populations. Calls for artificial propagation became most pronounced at times when abalone stocks wore thin. Finally, in 1963, California Department of Fish and Game (CDFG) biologist Keith Cox traveled to Japan to learn about that country's abalone hatcheries. Then, in 1970, the department started its own shellfish hatchery at Granite Canyon lab south of Carmel with the joint aims of rebuilding southern California's abalone populations and developing techniques that could be used by the fledgling aquaculture industry. Meanwhile, commercial divers also experimented on their own with growing and planting out "seed" abalones.[1]

Abalone mariculture offered a welcome, less-contentious alternative to ratcheting down fishing pressure, but it remained highly experimental. Although the CDFG had decades of experience with fish hatcheries intended to prop up depleted salmon and trout fisheries, there had been no reliable way to induce large numbers of abalone to spawn.[2]

That changed in the late 1970s when UC Santa Barbara molecular biologists Drs. Daniel and Aileen Morse began to study the biochemical processes controlling abalone spawning and larval settlement.[3] The Morses first discovered that the cue for synchronous spawning was a prostaglandin-like molecule. They determined they could coax abalones to secrete that molecule and spawn by mimicking a critical precursor enzyme with a small dose of readily available hydrogen peroxide. However, the resulting larval animals stalled out in their development after about a week. The Morses surmised there was some critical ingredient present in the natural environment—but missing in the lab—that triggered the larvae to settle and continue their growth.

With the help of marine biologist colleagues, the Morses began to explore the rocky reef habitats where the tiniest abalone dwelled. They ultimately discerned that larvae settled out in response to a chemosensory signal associated with crustose coralline algae, a red alga that looks like a coating of rosy pink paint splashed on low intertidal and subtidal rocks. When larval abalone touched down on coralline algae, they encountered an amino acid similar to the neurotransmitter GABA (gamma-aminobutyric acid) that prompted them to settle, metamorphose, and begin to grow into sedentary, shelled animals. The tiny abalone then fed on secretions of the algae until they developed the mouthparts to tackle hardier kelps.[4] It was a remarkable symbiotic relationship. Published in *Science*, the Morses' groundbreaking research cast new light on age-old, evolutionary-ecological relationships at a molecular level.[5] Their discoveries were also a boon to private aquaculture firms that began to spring up along the coast and would later become critical to wild abalone conservation efforts.

Although restoring abalone by planting out "seed" had been one of the most popular measures recommended by the Burge Report, implementation foundered because the legislature neglected to fund it. However, when mitigation money from Pacific Gas and Electric became available after a copper spill at its recently built Diablo Canyon nuclear power plant killed off hundreds of abalone, California Sea Grant (a federal program to fund marine research through state universities) and the Department of Fish and Game jointly launched, in 1977, an intensive three-year effort—the Experimental Abalone Enhancement Program—and hired Scripps Institution of Oceanography researcher Mia Tegner to head it up.[6] That choice would prove consequential.

* * *

Tegner had started her scientific career in the late 1960s in molecular biology, studying biochemical interactions between the eggs and sperm of sea urchins at UC San Diego. Finishing her PhD dissertation in 1974, she stuck with urchins but shifted to the field of ecology, "a gutsy move for a young academic," as one colleague later put it.[7] A field ecology class at Scripps introduced her to diving, and the experience changed her life. Tegner quickly became competent and comfortable with underwater research, and southern California's kelp forests beckoned with all manner of unanswered questions.

By the early 1970s, aiming to rein in the scourge of expanding urchin barrens, the state had started a kelp forest restoration program in southern California, hiring divers to kill the voracious kelp eaters (by smashing them with hammers and spreading quicklime) and then to transplant kelps.[8] Tegner worked on some of the early kelp transplant projects. The state also opened a new commercial red urchin fishery in 1972, tapping the Japanese market for *uni* (sea urchin roe) as another way to rebalance southern California's undersea ecology.[9] When Scripps ecologist Dr. Paul Dayton received a grant to research red urchins, he hired Tegner to help as a postdoc and urchin expert.

Starting in 1974, Tegner and Dayton dove in the Point Loma kelp forest off San Diego looking for juvenile red urchins, trying to figure out where the spiny animals settled and grew in their earliest stages. Juvenile purple urchins inhabited a wide variety of sheltering habitats, but Tegner and Dayton observed that baby red urchins nestled primarily into the tiny, protected space underneath the spine canopy of their elders. In an ocean full of hungry predators, this space was unexpectedly and disproportionately important.

The fact that baby red urchins appeared to be so tightly dependent on adults raised the possibility that the new commercial fishery could impair recruitment of the next generation. To test their thinking, Tegner and Dayton simulated the effect of fishing. They removed all legal-sized urchins on two large rock pinnacles and designated a third as a control where they let the large red urchins stay. After several months, including a period of reproduction, they found far fewer small urchins settled on the "fished" knobs and hypothesized that the spine canopy functioned as a nursery, affording both protection and food for baby red urchins, which they often saw eating from drift kelp snared and anchored by the adults.

They were surprised to also find baby abalones under the spine canopy of the urchins. Dayton would later recall finding nineteen little abs under the spine canopy of a single, old red urchin.[10] The adult urchins' forbidding spines seemed to safeguard the juvenile urchins and abalones from predation by sea stars, lobsters, sheephead, and other fishes until they grew large enough for their own shells to protect them. Tegner and Dayton published their findings in *Science* in 1977—underscoring how fishing could affect undersea ecological communities in unforeseen ways.[11]

Seeing baby abalones clustered under the large red urchin spines became Tegner's introduction to the complex interactions between urchins and abalone. At the time, the two animals were considered competitors for food and space. Earlier observations and research suggested that the decline of abalone had opened the way for urchins to take over. In severe cases, urchins formed fronts, aggressively devouring kelps and leaving abalone to starve. But the observation that red urchins also provided nursery habitat revealed a fascinating secondary relationship.[12]

Tegner wanted to better understand how the interspecies competition played out, so she designed a lab experiment, feeding kelp to abalones and urchins together in tanks. Under conditions of scarcity and food stress, she found that abalones hunkered down and waited for kelp to drift by, but the hungry urchins shifted from their usual sedentary habits and started to move and aggressively graze on the kelp—the behavior responsible for creating barren zones and preventing the reestablishment of a new kelp forest.[13]

Tegner concluded that, although competition between abalone and urchins may have been a key dynamic in the past, heavy fishing of urchin predators, such as sheephead and spiny lobsters, plus environmental degradation that diminished kelps, were now the dominant factors shaping southern California's kelp forest ecosystems.[14]

* * *

Though Mia Tegner had no direct experience in mariculture, her experience as a scientific diver working in kelp forests positioned her well to head up the three-year Experimental Abalone Enhancement Program focused on the field aspects of restoration.

Continuing to work from her base at Scripps, Tegner devoted herself to learning about abalone and was surprised to discover many critical gaps in fundamental scientific understanding. Previous researchers had studied food preferences, growth, hybridization, and reproduction with an eye to aquaculture, but still, too little was known about the mollusks' life history, recruitment success, predators, and mortality rates in wild populations.[15]

To evaluate the possibilities for abalone restoration, Tegner planned to test two approaches: outplanting hatchery-reared juveniles and translocating wild adult abalones that could reproduce in situ to repopulate depleted areas. She would also study the ecology of juveniles and the roles of abalone predators and competitors.

The Palos Verdes Peninsula was an ideal place to experiment. Its reefs—once abundant with pink, green, and red abalone—were now severely depleted. Lush kelp beds had been diminished by pollution from the sewage outfall near White's Point, and then were destroyed by a series of devastating storms in the late 1950s. The kelp forest had been gone for almost two decades, but with improvements in sewage treatment, the department's kelp restoration efforts of the late 1960s and early 1970s, and favorable ocean conditions, the undersea forest was finally beginning to rebound.[16] Still, there was no sign of abalone recovery. In 1976, the legislature closed the area to all abalone harvest, but it remained unclear whether populations could rebuild and recover on their own since there was so little broodstock.

In the spring of 1981, Tegner launched one of several experiments to test outplanting. Her research team of CDFG and Scripps divers cruised north from San Pedro, ferrying ice chests filled with hundreds of tiny farm-raised red abalone. At the study site, the biologists outplanted nearly seven hundred into the empty habitat and then tracked their fate at monthly intervals. A preliminary experiment had shown that the small mollusks more or less disappeared, likely lost to predation. So this time they removed all predators from the test plot ahead of time and put out slightly larger abalones in two size classes—about 2 and 3 inches—aiming to determine the best size for outplanting. They notched shells with a grinding tool for identification and then hand-planted the small abalone as best they could into protective habitat—deep crevices, the undersides of rocks, and the centers of boulder piles—hoping to give them the best shot at survival.[17]

Soon after Tegner's team started to plant out abalone, they confronted what would turn out to be their project's greatest obstacle. Within fifteen minutes, an octopus showed up and skulked, waiting—likely attracted by scent to the vulnerable young shellfish. The next day, Tegner caught an octopus eating two of the seeded abs and found forty Kellet's whelks—typically regarded as scavengers—roaming the newly planted area and engulfing small, stressed juveniles. Every month thereafter, they found a formidable cast of predators dining on their outplanted abalone. During a night dive, they found a spiny lobster prying one from the underside of a rock. In October, they pulled an octopus off another, and in December, they dove down to find three large octopuses hovering within two feet of a seeded abalone cluster. Through the course of the experiment, they would remove thousands of whelks, hundreds of sea stars, and dozens of octopuses.[18] The rocky nearshore habitat off the Palos Verdes Peninsula was a snake pit of predators.

Tegner quickly became expert at conducting abalone post-mortems. By looking at the shell remains they collected as evidence, she could readily determine the culpable predator. Chipped edges were the sign of hungry rock crabs or lobsters, and beveled drill holes meant octopus. The researchers noted the abalones' size at death, aiming to learn more about when their shells became large enough to provide "refuge."[19]

Tegner also found that many of the juvenile abs moved from the sites where they were initially planted into tight new clusters, suggesting either "gregarious behavior" or "agreement on preferred habitat."[20]

After a year of monitoring, the researchers "destructively sampled" the entire 25-by-25-meter study plot, which means they overturned all the rocks to search for the notched hatchery juveniles. They also looked outside the plot boundaries for evidence of outmigration. Of the 674 seed abalones they'd planted within the site, they found only eight—all hidden under sheltering boulders. The one-year survival rate was less than 1 percent.[21]

Department of Fish and Game biologist Kristine Henderson (now Kristine Barsky), who worked with Tegner on this project, would later recall they'd joked about feeling "like cocktail waitresses" serving hors d'oeuvres to octopuses while the state footed the bill.[22] And the bill was considerable. Though seed abalones cost only about 44 cents each, it took an estimated one thousand "seeds" to produce one adult—a cost of more than $400 per abalone.[23] Moreover, Tegner observed that hatchery-raised juveniles lacked "street smarts." They'd move out into the open during the day, expecting to be fed, and instead become lunch for an octopus. In contrast, wild abalone that grew up on reefs hunkered tight in protected crevices during the day and foraged only at night.[24] Hatchery-raised juveniles typically looked different, too: in one case, brighter green owing to diet; and in another, gleaming because flowing water in propagation tanks had eroded the small shells to expose their iridescent nacre—in both cases, likely drawing lethal attention from predators.

When Tegner compared her findings with more promising results reported from Japan, she realized that Japan didn't have predation problems, in large part because it had more aggressive fisheries that specifically targeted octopus and other abalone predators. In California, fisheries had the opposite effect, significantly reducing populations of octopus predators such as moray eels—heavily fished for an Asian live-fish market—thus leaving the eight-legged abalone eaters to abound. Meanwhile, heavy fishing of spiny lobsters allowed their prey—Kellet's whelks—to also proliferate, with likely impacts on juvenile abalone survival.[25]

Through the early 1980s, Tegner and other researchers at UC Santa Barbara conducted several additional, larger-scale outplanting experiments—seeding out about a hundred thousand juvenile abalone from public and private aquaculture facilities to reef locations on southern California's mainland and islands.[26] When Tegner reviewed the results of all these experiments and found similarly discouraging results (the one-year survival rate for all the experiments was less than 3 percent), she concluded that outplanting abalone in southern California for fishery enhancement was futile.[27]

* * *

Tegner also set about testing the premise underlying the need for a restoration program. She wanted to know whether existing abalone stocks could rebuild if left alone—in particular, whether green abalone populations still present on the Channel Islands could naturally "reseed" severely depleted reefs along the Palos Verdes Peninsula. Tegner used drift tubes as a surrogate to study how currents transported abalone larvae to assess whether those from the islands could make it all the way to the mainland, and if they could do so within the narrow five-day period that the minuscule animals have to find a promising place to settle.

In June 1981, she let loose the first set of 1,200 small glass tubes from areas known to harbor green abalone on both the Channel Islands and along the mainland.[28] According to previous lab studies, green abs spawned in early summer and also in the fall, so she set loose another set of drift tubes in late October.[29] The tubes were weighted to float in the upper water column, where abalone larvae swim. Cards inside exhorted beachcombers to provide information about where and when each tube was found and to return cards by mail.[30]

The following winter when Tegner mapped the data, she found that only four of the thousands of tubes released turned up on the mainland within the critical five-day larval lifespan, and of these, only two had landed in suitable green abalone habitat. Though Channel Islands abalone could occasionally contribute genetic diversity to mainland populations over an evolutionary time frame, the drift tube experiment made it starkly clear that island broodstock could not naturally supply enough larvae to reseed depleted areas on the mainland. The fishing closure enacted by the legislature in 1977 would not be enough to restore abalone to the Palos Verdes Peninsula.[31]

The drift tube study also intimated something new. Many of the tubes had moved only a very short distance from where they were released, suggesting that abalone larvae didn't travel far from their parents. To have any hope of rehabilitating the fishery, moving adults that could reproduce on their own now looked to be the most promising way to replenish empty areas.[32]

That November, the research team started to test this possibility, too. Tegner and her CDFG colleagues began collecting mature green abalone from the windward side of Santa Barbara Island and moving them to empty habitat on the Palos Verdes Peninsula. Over the course of ten months, they moved 4,453 mature animals. One researcher graded the abalones' reproductive potential by visually examining the bulk of their gonads—packed tight under the shell beside gills and digestive tract. Bigger was better in terms of reproductive readiness. Then, divers carefully placed each animal in a favorable spot, clustered close so that clouds of sperm and eggs might readily mingle and mix. At monthly intervals, the researchers dove down to check on the transplants, specifically noting the condition of their gonads and actively removing predators.[33]

The next summer, they observed a telltale increase, and then drop, in gonad bulk—a sign of spawning—and were hopeful. "They were growing like gangbusters, and they were reproducing," Tegner would later report.[34] But in that fall of 1982, El Niño conditions brought warm water and intense storms that ripped apart nearby kelp beds, leaving the transplanted animals with little food from January through May. The following year, the green abalone showed little or no growth. In the spring of 1984, their gonads remained small, and there had been no more spawning.[35]

Then in the spring of 1985, nearly four years after the initial transplant, Tegner and her colleagues finally found a positive sign—a pulse of newly emerged juveniles. The transplanted green abalones had successfully reproduced just before the El Niño, and the tiny juveniles had managed to persist through warm water and storms.

But the promise of the new recruits was soon dashed. "Basically, the adults, all tagged, disappeared," Tegner would later recall with frustration. In forty-five dives, they could find only 9 percent of the green abalone they'd transplanted.[36] Because there was no evidence of masticated or drilled shells, she surmised that the missing abalone had been taken by poachers. With no dedicated warden's boat for the area, the Department of Fish and Game patrolled in the vicinity only about twenty-four hours out of every month—not enough to deter illegal activity.[37] Without adequate enforcement, the fishery closure that everyone had hoped could bring abalone back to the Palos Verdes Peninsula would not succeed.

The experiment showed that transplanting adult broodstock could work to restore abalone in depleted areas, but the strategy was ultimately limited by the still-dwindling supply of large adult animals and also by irrepressible human appetites. This was not the only transplant effort scuttled by poachers. For the pragmatic Tegner, the poaching incidents raised a sobering question: Was restoring abalone actually "doable" in southern California "where there are 15 to 18 million people within a stone's throw of the coast"?[38]

The benefits of transplanting could accrue only after adult animals had the chance to spawn over the course of several years, and although CDFG biologists and fishery managers had long considered abalone to be highly prolific, Tegner's research spotlighted the fact that a range of factors—from predators to warm water influxes to poachers—now routinely interfered with the animals' reproductive success.

The results of Dr. Tegner's research in the late 1970s and early 1980s convinced her that the state's wishful projects of outplanting juveniles—and even translocating adults—to restore southern California's abalone fishery were not only infeasible but also fundamentally misguided, especially because optimistic expectations for these experimental efforts had been used to justify deferring the much-needed commercial fishing restrictions that Burge and his team had urged.[39]

Despite Tegner's unequivocal findings and additional experiments conducted through the 1980s that showed only marginally better results when juveniles were planted out into protective structures, the promise of seeding out abalone would continue to hold a legion of wishful adherents among enterprising mariculturists, commercial abalone divers, and sea otter advocates—all seeking to solve the otherwise intractable problem of too few abalone.

Rooted deep in political dynamics that made it seem impossible to restrain fishing pressure, the Abalone Enhancement Program followed the hopeful belief that human ingenuity could fix environmental problems and restore damage. However, in this case, Mia Tegner's ecology-informed field research showed that the prospects for artificially enhancing the abalone fishery were poor at best. As she explained, "We are left with the more painful prospect of carefully managing the abalone stocks that remain."[40]

PART 4

Crisis
(1980–1997)

Past experience has shown us that opposition from the commercial abalone industry will attempt to prevent the legislature from enacting legislation to protect the abalone resource for the future. This short-sightedness is destroying California abalone populations, a wildlife trust that belongs to all the people of California.

> —Paul Turnbull, Abalone and Marine Resources Council, to
> Robert Treanor, October 10, 1996, California State Archives

So where do divers fit into this and why should we care? The reason, simply put, is that if we don't take the initiative in caring for our seas, then nobody will.

> —Steve Benavides, Catalina Conservancy Divers, guest
> op, news clipping in California State Archives

I'd like my kids to be able to get an abalone someday—or at least see one, but if something's not done soon, that'll never happen.

> —David Woodland, captain of sport-diving boat,
> *San Jose Mercury News*, March 25, 1997

It is difficult to get a man to understand something, when his salary depends on his not understanding it.

> —Upton Sinclair, *I, Candidate for Governor*, 1934

We felt that we were protecting the marine resources for the people. I always felt that the public interest should carry more weight than the commercial interests.

> —Pete Haaker, California Department of Fish and Game,
> retired marine biologist, interview, April 2010

CHAPTER 10
Losing Grip

Sheltered from the rowdy blow of northwesterly winds, Johnson's Lee is a popular moorage located off the south shore of Santa Rosa Island—third from the coast in the northern Channel Islands chain. Sharp winds blowing past the island stoke strong currents that stir up a steady upwelling of cool, nutrient-rich water that nourishes a robust stand of giant kelp in the lee. And about thirty feet beneath the water's surface, a large reef harbored an abundance of red abalone, long considered a "very good bed" by commercial abalone divers.[1]

Its plentiful abalone and isolated location also made Johnson's Lee an ideal place for research. One of the Burge Report's key recommendations was for the state to initiate growth and life history studies for the various abalone species in southern California. Although the state's management of abalone had long been based largely on size limits, there had never been any analysis to determine whether the proscribed limits actually worked. To follow through, Dick Burge had designed a five-year research program to collect data about the animals' growth and fecundity—key parameters to determine how much fishing could be sustained.[2]

Although the California Department of Fish and Game (CDFG) had cut funding for abalone research, Burge, with Mia Tegner's help, managed to tap some Pacific Gas and Electric Company–Diablo Canyon mitigation funds. Together, with additional monies from Sea Grant, they planned to conduct joint studies to fill in critical gaps in fundamental knowledge about abalone. They recruited dozens of divers from California's scientific diving community—including biologists from CDFG, Scripps, Moss Landing Marine Labs, and UC Santa Barbara—chartered a dive boat from the Santa Barbara–based dive-tour company Truth Aquatics, and, starting in 1978, organized dive teams to head out to Johnson's Lee each July for a week of intensive research. After Burge left the department in 1980, Tegner and other department biologists, including Pete Haaker, who would later become the state's go-to guy for abalone, carried the research forward.[3]

The divers set thirty-meter transect tapes along the bottom and then tagged abalone along those lines. The following year, they'd recapture the mollusks and note their length and location, the presence of juveniles, and evidence of any dead animals.

Researchers could wring a surprising amount of information from such basic data, especially with recent advances in computer analysis.[4] Burge and Tegner wanted to analyze growth rates not only in a general sense but also for abalone of different sizes and living in different habitat types. They wanted to track the animals' density and to get a better sense of where juveniles lived, what predation pressures they faced, and mortality rates by size. Through five years of research, forty-one divers tagged and successfully tracked more than 2,100 animals—the largest fishery-independent abalone study to date.[5]

Since Johnson's Lee remained a popular fishing site, the biologists found few legal-sized abalone, only three in one hundred. A sharp drop off in numbers within 5 mm of legal harvest size indicated that pry-bar cutting by divers continued to be a problem. Despite the 1976 limited-entry legislation that had aimed to curtail this problem, the researchers estimated that mortality from pry-bar cuts hovered at 10 percent, a rate that could still damage the fishery.[6]

The biologists also determined that growth rates were not only variable, but highly variable, even in the same area. It could take anywhere from five to eleven years for a red abalone at Johnson's Lee to reach legal size for sport harvest and from seven to nineteen years to reach commercial harvest size—a realization that cast further doubt on the prospect of using traditional fisheries models, which assumed growth rates to be constant across entire regions, to manage abalone.[7]

Most important, the researchers discovered that abalones' success at reproduction could also vary wildly.

* * *

The biologists thought that the 1982 dive trip would be their last. Then came an influx of warm water, followed by a series of severe fall and winter storms with massive and destructive waves that ripped gangly kelps from their sturdy holdfasts off the mainland and islands. Unusual subtropical species, such as delicate paper nautilus and mantis shrimp, appeared, and the sea surface temperature lingered above 68 degrees Fahrenheit for more than ninety-three days straight. Even waters to a depth of sixty feet warmed. It would turn out to be the strongest El Niño on record thus far in California.[8]

Although previous observers had noticed that southerly fishes occasionally moved north into California with pulses of warm water, not until the 1970s had scientists finally recognized that the recurrent oceanographic phenomenon known in Peru as El Niño was closely related to larger atmospheric cycles. And it wasn't until the winter of 1982–1983 that scientists recognized that these same large-scale oceanographic oscillations gave rise to the warm water and storms that had episodically hit the Northern Hemisphere—what they started to call El Niño in California, too.[9]

Most important, they began to recognize that El Niño's warm water and storms weren't just fluke events but rather inevitable occurrences that disrupted marine ecosystems on a periodic, albeit sporadic, basis.

For biologists studying California's abalone, the 1982–1983 El Niño created the opportunity for an experiment. The information they'd carefully collected in the late 1970s and early 1980s was now an ideal set of "before" data. They decided to return to Johnson's Lee one more time to repeat transect surveys, recapture tagged abalone, and see firsthand what changes El Niño had wrought underwater.

When the scientist divers submerged once again, they found the lush kelp forest they'd come to know reduced to tatters. Red abalones still hunkered on the reef, but there were few animals above the sport minimum size.[10] When the researchers compared the data from before and after the El Niño, they determined that abalone growth was severely depressed and no successful recruitment had occurred for two years. Mia Tegner, who was also working at Palos Verdes Peninsula and Point Loma during this El Niño period, reported similar findings at those sites, too.[11]

Biologists already knew that red abalone lived only in the coolest recesses of southern California's underwater seascape and that red abalone larvae grew best at temperatures below 68 degrees Fahrenheit, but the El Niño showed that warm water could hinder both spawning by adults and survival of larvae, revealing that the animal's reproductive success was intimately and unequivocally tied to variations in seawater temperature.[12]

The full repercussions of the 1982–1983 El Niño for red abalone's reproductive success wouldn't become clear until years later, but the harvest at Johnson's Lee soon resumed. By 1987, abalone growth had picked back up, shellfish reached legal size, and commercial divers reported that the reefs there were again productive and ripe for harvest.[13]

The 1982–1983 El Niño was a fluid echo of the warm water influx of 1959. Mia Tegner and Pete Haaker would later analyze and compare both events. In both cases, abalone failed to reproduce for several years in a row, leaving whole age classes missing from the up-and-coming population. This created an unforeseen gap in the population of fishable shellfish ten to twelve years later. Meanwhile as existing adults grew to legal size, they were fished out, leaving no parent broodstock to reproduce. While abalone could withstand elevated temperatures and poor food for a while, added fishing pressure made population reductions "almost inevitable."[14]

Since the days of Scofield, the California Department of Fish and Game and commercial divers had believed that size limits served as a kind of refuge that could assure the persistence of fishable shellfish populations, based on the assumptions that abalone regularly spawned, reproduced, and grew; but changing conditions both in the structure of abalone populations and in the environment showed those assumptions to be deeply flawed.

When Pete Haaker joined the Johnson's Lee study in 1980, he had no idea that he'd ultimately be the one to write up the data, that he'd end up a decade later as California's lead abalone biologist, or that managing the state's abalone fishery would become so complex and urgent. When years later I asked Pete about the studies at Johnson's Lee, he explained, with the benefit of hindsight, "We know now that warm water affects kelp productivity so, for abalone, El Niños have a metabolic effect, a food effect, and a reproductive effect. If you keep fishing hard at the same time you have unsuccessful reproduction and settlement, you can really damage a population."[15]

In 1993, Haaker would return to Johnson's Lee. Where researchers had found 632 abalones per hectare in the years before the 1982–1983 El Niño, he would find only sixty.[16] When Haaker returned to dive at Johnson's Lee in 1995, he'd find only six animals.[17] And when Haaker's CDFG colleague and successor Ian Taniguchi conducted five swim surveys of the same area in 2001, he'd find none.[18]

More pieces of the puzzle were coming together. The return of sea otters to the Central Coast in the 1950s and 1960s had offered a crucial clue to understanding how ecological interactions among living creatures in kelp forests could affect abalone. Department studies of fishing pressure in the 1970s had documented how relentless take had significantly whittled down abalone broodstock. Then, the 1982–1983 El Niño revealed how episodic periods of drastic environmental change—elevated temperature and massive storm waves—could affect abalone growth and reproduction, with the consequential result of further reducing populations.

For abalone, it was one stress and then another—fishing pressure, water pollution, El Niño—hitting like battering waves—all eroding the legendary mollusks' capacity to endure. But it would take an even greater crisis to refocus attention on the plight of California's abalone.

* * *

In the early 1980s, while radios blared Dire Straits on the mainland beaches, out at the recently created Channel Islands National Park, biologist Gary Davis was setting up a new program for ecological monitoring. A tall, broad-shouldered man, Davis had grown up diving in southern California, studied marine biology, and then moved east in the late 1960s to start his park service career in the Everglades. There, seeing the grave impacts of water pollution on the marine life of Florida Bay, he'd come to believe that national parks needed baseline ecological studies and then ongoing monitoring so managers could better understand changes and more readily make science-based decisions to safeguard wildlife.

Establishing a baseline was his goal. The initial challenge was to pick a set of indicator species for each ecosystem type. Black abalones were a natural choice for the intertidal zone. The dome-shaped mollusks covered the nearshore rocks in some places like paving stones, far more plentiful on the Channel Islands than on the mainland coast, where they'd long since been plucked off by shore-pickers. Gary hired Dan Richards, a lanky young marine biologist just out of Humboldt State, to help with the survey and monitoring work.

Pete Haaker, the California Department of Fish and Game marine biologist who'd worked on kelp restoration projects off the Palos Verdes Peninsula, and with abalone research at Johnson's Lee, also joined the National Park Service biologists in their Channel Islands surveys. The interagency collaboration was uncommon at the time—the result of an awkward split of jurisdiction. Channel Islands National Park had started first as a national monument in 1938, made up of only Anacapa and Santa Barbara Islands and managed by the distant Sequoia National Park. In the 1950s, the federal government gave California jurisdiction over its "submerged lands" for the purpose of oil and gas drilling. Meanwhile, as oceangoing boats improved, fishing pressures around the remote islands increased. By 1966, when national park rangers first started living on Anacapa, the islands' waters sustained an active fishery with a sizable commercial fleet plus recreational "party boats." The early rangers became concerned that heavy fishing was damaging the park's marine ecosystems and urged closing half the islands' shorelines to fishing.

However, with fish, lobster, and abalone populations drastically declining on the mainland, commercial fishermen pushed back, pressuring the state to reexert its authority over "submerged lands" and to open up all the islands' nearshore waters to fishing. The fishing industry appealed the case all the way to the US Supreme Court, which ultimately gave California full jurisdiction over all the monument's waters in 1978. To this, Congress responded in 1980 by expanding the monument into a full-fledged national park, adding San

Miguel, Santa Cruz, and Santa Rosa Islands, including a one-mile ribbon of submerged lands around each one, and explicitly directing the park to monitor conditions and make recommendations to protect park resources. In effect, this meant the park and the state would need to jointly manage the new park's marine resources, including fisheries. The model in the Channel Islands would be to collect one credible data set that could be used by both public agencies.[19]

To Gary, Dan, and Pete, the mandated collaboration was amenable. The three biologists worked together a lot those first summers, running the national park's research vessel out to the islands and lingering for weeks at a time. Their hair bleached, their skin tanned, and their fingers calloused as they set up research plots and then began to gather baseline data—counting mussels, snails, crabs, lobsters, sea stars, and abalone from the islands' beaches and intertidal zones down into the rocky depths of surrounding kelp forests.[20]

In their first year, 1981, they routinely counted one hundred black abalones per square meter, Gary told me. Imagine a card table piled high with one hundred abalones! At low tide, black abalone spread everywhere, leaving little space for anything else. Like all intertidal animals, the black abs were tough, withstanding the full force of waves at high tides and then exposure and desiccation during low tides. The animal's muscular foot got a vigorous workout holding tight in the crashing waves of the surf zone, and its strong shell

10.1. In 1983, intertidal black abalone remained abundant at Channel Islands National Park, shown here at Santa Rosa Island. However, by the 1990s, an epidemic of withering syndrome had wiped out nearly 99 percent of these animals in southern California, raising concern that other abalone species could be vulnerable. (Photo by Gary Davis)

10.2. Marine biologist Gary Davis led the effort to set up an ecological monitoring program at Channel Islands National Park, where he and his colleagues documented how withering syndrome devastated black abalone populations. (Photo by Dan Richards)

clamped hard against the rock, protecting it from drying winds and predators at low tide. In the dead of winter, when winds chilled the air below freezing on the northern islands, the biologists found black abs with ice crystals on their feet. In summer, they found black abs enduring great heat on the dark rocks when exposed to direct sun by low tide. One sweltering day, out of curiosity, Dan inserted a thermometer under the foot of a black abalone and took a reading of 95 degrees F—surprisingly high for a "cold-blooded" invertebrate.[21]

Given the black abalone's abundance and hardiness, the biologists were not at all prepared for what they found when they revisited their plots in 1985—abalone by the hundreds hanging limp, barely able to hold. An abalone's big-footed body usually fills its shell, but many of these animals were atrophied, in some cases, shrunk down to half size. The fleshy epipodial fringe that extends in tentacles beyond the animals' shells was not its signature jet black but rather pallid gray. As the biologists surveyed their plots on Anacapa Island, abalone fell loose, one after another, shells crashing onto rocks below. Typically, an abalone removed from its rock will immediately and vigorously right itself to regain the protection of its shell. But these animals seemed almost comatose.

Doing some quick math, Gary determined that the mysterious wasting was affecting 10 to 15 percent of the abalone in their plots and beyond. They continued making their counts and measurements and found the same

situation when they went to Santa Rosa Island two months later, though abalone at the other islands remained healthy. By the time they returned to Anacapa in the fall for the next round of monitoring, even more abalone looked
sick, and many had died. In just twelve months, the island's population had
declined by 30 percent.[22]

Soon they found windrows of shells, thousands of them, piling up on
the beach in bright silvery lines—beautiful but eerie, like finding so many
skeletons. That's what cued them to realize that what they were observing was
not some minor population fluctuation: it was something much bigger.

As the months went on, they found the sickly abalone at more sites and
on more islands. Fishermen and other scientists reported the mysterious condition, too. It seemed to be spreading slowly from the south to the north sides
of the islands and from the islands closest to the mainland to those farther out.
In his official year-end monitoring report for 1987, Dan described the black
abalone decline on the northern Channel Islands as "alarming."[23]

✳ ✳ ✳

Anyone observing the intricate patchwork of the intertidal zone becomes
acutely aware that every creature has a different approach to life. Blue mussels
hang tight, stitched to rocks with golden byssal threads, and strain plankton
out of seawater. Leggy sea stars slowly roam, pry open the anchored mussels,
and then evert their mouths to suck out the bivalves' meaty contents. Crabs
skitter and scavenge, and chitons and snails graze, while black oystercatchers
stride with pink-stockinged legs, probing with orange bills for tiny crustaceans,
baby mollusks, and other creatures plying amid rugged ranges of barnacles.

To grasp the severity of the mystifying condition that the biologists discovered, remember that an abalone's modus operandi is attachment. With its
large muscular foot, it grips onto its rocky home spot and mostly sits, waiting
for pieces of kelp to drift close enough to grab and pull to its mouth. Though
young abalone may amble around to forage under cover of darkness, once an
animal finds a sweet home spot, it may not budge for months, or even years if
an adequate supply of kelp drifts by. Pulling its shell tight against the rock, the
animal has protection against the pounding of the surf and its predators. It is
said that a 6-inch abalone can hold tight with four hundred pounds of force.[24]
But the abalone's strong foot and tough shell were no match for whatever was
cutting them down now.

An abalone that loses its grip becomes utterly helpless, like a bird that
can't fly or a fish that can't swim. Some floated like forlorn little craft, buffeted

by waves. Dan noticed that gulls would occasionally peck at the belly-up black abs, but most predators avoided eating the sick animals.

✳ ✳ ✳

The biologists had dutifully tried to establish a baseline for black abalone, yet here it was already changing radically before their eyes. They soon realized they'd need to tap other types of scientific expertise to understand what was causing the massive die-off.

One possibility was starvation. Similar withering had been observed decades earlier on the Palos Verdes Peninsula after big storms in 1958 devastated kelp forests. At that time, when the withered abs were transplanted to kelp-rich areas, they recovered. The recent 1982–1983 El Niño storms had once again sharply reduced the supply of drift kelp. To test the starvation hypothesis, the biologists sent fifty-two black abalones—some healthy, some emaciated—to the Cabrillo Marine Aquarium in San Pedro. All were placed in seawater tanks and amply fed. Yet within a month, all sixteen of the feeble abalones and seventeen of the apparently healthy animals had died. Starvation was not the problem.[25]

The animals' sickly appearance strongly suggested the possibility of poisoning or disease. The biologists sent withered black abs to the CDFG's pesticide lab, but toxicologists found no evidence of toxins and detected only insignificant amounts of heavy metals. The state's fish disease lab found no problematic pathogen.[26] Lacking an identifiable "smoking gun," the biologists started to call the mysterious killer "withering syndrome."

An important clue came in June 1988, when withered abalone were detected for the first time on the mainland near the Diablo Canyon nuclear power plant, about a hundred miles north up the coast from Point Conception. With the plant's construction in the early 1980s, consulting biologists had started to monitor the impacts of its warm water effluent on the ecology of Diablo Cove.

In the power plant's first three years of operation, they'd found no particular effects on the cove's black abalone, but then the mysterious withering hit like a plague. By 1989, 90 percent of the abalones were dead. Mortality was highest in areas closest to the power plant's outflow, where temperatures were highest.[27] This fit with Gary, Dan, and Pete's field observations that withering syndrome had first appeared and then developed fast and furious on the islands' warmer south sides and then more gradually proceeded to decimate populations on the cooler northern shores.

The die-off at Diablo Cove pointed to a new hypothesis. It seemed that the stress of elevated temperature was making animals more vulnerable to some yet-to-be detected pathogen or parasite. As the lethal condition continued to spread, they worried and wondered, what could it be?

∗ ∗ ∗

From the time Gary, Dan, and Pete saw the first sickly animals in 1986, they watched year after year as the northern Channel Islands' massive intertidal populations of black abalone literally withered up, died, and washed away. By 1991, 99 percent of the black abalones on Anacapa, Santa Barbara, and Santa Rosa Islands were gone. In 1991–1992, there was another El Niño, and by 1993, 90 percent of abalones initially present at all islands were gone. Only a very few sites at the most remote islands—San Miguel and San Nicolas—still harbored small clusters of healthy-looking black abalone. And as the populations dwindled, the possibility for new reproduction diminished markedly.[28]

For the biologists, the experience of witnessing the black abalone epidemic firsthand was profound. To see such a drastic transformation in the intertidal zone in such a short time heightened their awareness of how quickly conditions could change. After so many days spent observing, counting, measuring, and analyzing abalone—especially sick abalone that were dropping helplessly from the rocks—they felt protective.

What Gary, Dan, and Pete saw on the Channel Islands during those summers in the late 1980s fundamentally changed the way they saw abalone. What for most of their lives had been an abundant and delicious shellfish was fast becoming an imperiled animal.

Meanwhile, they started to learn about new research from across the Pacific that made the black abalone epidemic seem even more dire.

CHAPTER 11
From Crisis to Closure

Halfway round the globe, Australia had recognized the vulnerability of its own abalone early on. Before 1970, the government had instituted limited entry and tight landing quotas in its fishery and then focused research on basic biology, population dynamics, and sustainable management.[1] With marked success, the Australians were in a good position to offer advice about abalone. Starting in the mid-1980s, down under marine biologist Dr. Scoresby Shepherd began to travel regularly to Baja to consult with Mexican researchers.

As California's abalone landings dropped precipitously through the 1970s, so too did Mexico's. With commercial ab divers in Baja targeting the same species as southern California's fishery, and managers there following California's laissez-faire strategy, the Mexican fishery had declined to near collapse in 1981. That crisis had prompted the Australian government to fund a collaborative foreign-aid research effort aimed at restoring and better managing Mexico's abalone fishery.

En route to Baja, Shepherd flew through Los Angeles and soon met Mia Tegner at Scripps. They talked abalone, struck up a friendship, and in the years that followed, she occasionally drove across the border to help him with research dives off nearby Ensenada.

As Shepherd shared his own research and the findings of his Australian and New Zealand colleagues with his new California friends and Mexican collaborators, they all recognized there would be tremendous value in bringing abalone researchers from different countries together for a conference. This was long before the internet made it easy for international scientists to share their work. It took countless screechy faxes back and forth across the Pacific to organize the landmark gathering.[2]

This first International Symposium on Abalone, held in La Paz in 1989, would prove to be a milestone in abalone science, allowing for exchange of emerging knowledge at a crucial time. Mexican biologists hosted the event, while Shepherd invited researchers from the South Pacific and Tegner tapped West Coast biologists. All told, scientists from Australia, New Zealand, Japan, the United States, Canada, Mexico, the Middle East, Europe, and South Africa gathered to discuss the state of their respective abalone species and share their latest findings. All faced similar problems: declining stocks, disease, unresponsive management policies, recalcitrant fishers, limited budgets for research and enforcement, pernicious poaching, and the impossible expectation of high and sustained harvests for a slow-growing and increasingly valuable yet vulnerable animal.[3]

For the state and federal biologists working in California, the symposium put their own experiences with black abalone into fresh perspective and provided a broader biological framework to understand what was happening with other declining *Haliotis* species. In particular, Shepherd's long-term monitoring of Australian abalones had shown that reproductive success was tightly linked to the density of the mollusks. In fact, some Australian species aggregated each year to spawn. (No one even knew whether California abalone species aggregated to spawn; the animals had always been so prevalent.) Shepherd had also found that when the density of abalones dropped below a certain threshold, recruitment success in the following years became erratic and populations foundered.[4]

It was the first time that the "Allee effect" had been documented for a marine mollusk. Back in the 1930s, American wildlife biologist Dr. Warder Clyde Allee had found that, at low populations, animals' reproductive success declined because the fewer remaining individuals were unable to locate mates.[5] In the case of abalone, when populations dropped too low, leaving animals too far apart, there was simply no way for their gametes—broadcast into seawater—to meet up for fertilization. The Australian biologists had found this situation with several of their abalone species, underscoring the crucial importance of maintaining populations at a sufficient density to sustain successful reproduction.

Meanwhile, another Australian biologist, Dr. Jeremy Prince, had determined that the larvae of some abalone species traveled only a short distance before settling, affirming what Mia Tegner had found with her drift tube study and calling into question the long-standing assumption that abalone larvae routinely dispersed across broad distances. His research confirmed that depleted areas would not necessarily rebound naturally, even if fishing pressures were significantly reduced.[6]

These two findings had troubling ramifications for abalone in California. If density of adults dropped below the critical threshold and there was no possibility of natural recovery through settlement of larvae from elsewhere, the only hope was to conserve broodstock in localized areas—to save the wild abalone we still had. At once, it became eminently clear to Gary, Dan, and Pete that withering syndrome was already reducing black abalones in southern California below the critical threshold for reproductive success, leaving almost no local broodstock to rebuild populations. They realized other abalone species could be at high risk, too.

Following fast on the heels of the black abalone epidemic, the 1989 International Symposium on Abalone marked a turning point in the thinking of California's marine biologists.[7] They realized the state could no longer manage southern California's iconic abalone for fishing—it needed to do everything possible to help abalone survive.

* * *

As withering syndrome spread rapidly from island to island, California Department of Fish and Game (CDFG) and National Park Service (NPS) biologists knew that the still-open fishery for black abalone was also whittling down the remaining population. In the northern Channel Islands, where the water was colder, they'd found the disease was advancing more slowly, but their surveys indicated that fishermen were still taking animals from these very areas.

They knew the best hope for the future of the species lay in the possibility that some small pocket of remaining animals might be resistant to the still-unknown pathogen. But if all the healthiest black abalones continued to be taken by commercial fishermen out on the islands and by shore-pickers on the mainland, southern California's entire population could wink out in a very short time.[8]

While the biologists became increasingly alarmed, commercial divers—feeling pinched by reduced landings—began to push, once again, to open new areas to harvest, northward up the mainland coast, where the disease had yet to advance. Because black abalone had become a significant portion of the commercial catch after other species became scarce, dwindling populations in southern California now made it difficult for divers to keep up with the state's six-thousand-pound minimum-landing requirement.[9] Originally enacted as part of the 1976 abalone law to weed out less-experienced divers, the mandated annual catch stipulation now served counterproductively, compelling commercial divers to search out the rapidly disappearing animals to keep their permits in good standing.

To address this and other problems, the Department of Fish and Game worked together with commercial divers of the California Abalone Association (CAA) and North Coast representative Dan Hauser to forge new legislation—the first significant abalone statute since 1976. The 1990 law lowered minimum landing requirements to 1,200 pounds to account for the much-diminished abalone supply. It also, finally, lowered the limited-entry target to seventy commercial divers—what the Burge Report had recommended fifteen years earlier (124 permitted commercial divers remained)—and required that any new commercial diver buy out two existing permits to enter the fishery, the reductions still working by attrition only.

In response to some egregious North Coast poaching cases, the new law sharply increased penalties for illegal take and also established possession limits for commercial divers working the San Francisco and San Mateo coasts and the Farallon Islands, with the idea that ceiling caps would deter poaching (landing quotas had also been part of the Burge Report's 1976 recommendations). The new law also levied an additional landing tax that commercial divers would pay to fund abalone enhancement and restoration projects—outplanting juveniles and trying to seed out "spawn" (larval abalone), the approaches they most favored to sustain the fishery.[10]

Although the CAA pressed hard to open new areas of the Central Coast to black abalone harvest, the Department of Fish and Game rejected those proposals and negotiated the closure of three Channel Islands (Anacapa, Santa Cruz, and Santa Barbara) to harvesting of blacks, though by that time there were virtually no black abalone to be found on the mainland coast south of Point Conception, and island populations were already scant.

The new law was a stopgap measure at best. CDFG biologist Pete Haaker knew that to protect the species, the state would need to completely close the black abalone fishery as soon as possible. He urged his supervisors to make an emergency closure, but they told him he'd need to pursue a full-blown public process—with scoping meetings, hearings, and preparation of an environmental review to present to the Fish and Game Commission for a final decision. With potentially disease-resistant animals continually being fished out, Haaker thought the situation was too urgent for so much time-consuming paperwork and bureaucratic process, but, based on his experience hammering out the 1990 compromise law, he knew, too, that closing the black abalone fishery would be a political challenge.

With no alternative, Haaker and his department colleagues doubled down and got to work. They began to float the closure idea with citizen members of the state's Ad Hoc Abalone Advisory Committee (mostly recreational divers) and with commercial divers on the Director's Abalone Advisory Committee. Then they conducted an environmental review—analyzing the threat, considering options, and officially recommending closure of the black abalone fishery in 1993.[11] Their final report warned that the Fish and Game Commission would need to choose between "short term utilization of the fishery, probably leading to a total collapse of the resource, or closing the fishery in the hope that restoration will occur, either naturally or with human intervention."[12] To Haaker, a long-term view made the best choice clear.

But commercial divers knew places where healthy animals persisted—some out at the remote San Miguel Island and others north up the coast in the still-cool waters of central California. In response to the department's recommendation to close the black abalone fishery, the CAA persisted in pressing the Fish and Game Commission to open previously closed areas north of Point Conception. The commercial divers believed that, if all the black abalone were going to die anyway, the state should allow continued harvest of healthy animals for some economic benefit.[13] Moreover, they were irked and worried that translocation of sea otters to the remote San Nicolas Island in 1986—a plan hatched in the 1970s to move otters away from the Central Coast and start a second population—would bring added predatory competition for shellfish to southern California waters, and they wanted a hedge. The commercial divers were accustomed to a long-standing department calculus that offset imposed fishery losses with increased opportunities for take, either by reducing size limits or opening new grounds or new target species. That was the way it had always worked in the past.

But this time, Haaker and his colleagues carried a new understanding of the heightened vulnerability of California's abalone—and a deep resolve to protect them.

* * *

When the CDFG put out its draft environmental review recommending closure of the black abalone fishery in 1993, no ocean or wildlife conservation groups weighed in on the mollusk's fate; marine conservation had not yet emerged as an effective movement in California. Nor did any foragers or recreational divers; their interest in the now mostly vanished nearshore abalone had flagged. Only three people commented—all associated with the commercial industry, and all in opposition.

CAA president and diver John Colgate argued that a fishery closure would not stop the disease from spreading and that size restrictions were already in place to sustain healthy abalone populations.[14] CAA treasurer and diver Jim Marshall regarded the department's argument about conserving genetic diversity of black abalone with skepticism and worried that closing the fishery would end attention to the animals' plight and "relegate the species to a back shelf somewhere." He thought continued commercial fishing would best "ensure the flow of information on the WS [withering syndrome] problem" from fishermen to research biologists.[15] Marshall also questioned the data used to support the closure because it had been collected, in part, by biologists from the National Park Service—an agency whose mandate, he argued, was "to remove commercial interests" from within park boundaries.[16] Finally, Steve Rebuck, whose dad had worked as a commercial diver during the abalone fishery's late-1950s heyday and who'd been hired by divers as their advocate, resisted the closure, instead urging the commission to work toward continued cooperation to "find the source of the problem."[17]

When the California Fish and Game Commission convened in Sacramento that June, the department's deputy director, Al Petrovich, presented his agency's recommendation to close the black abalone fishery. He explained that the CDFG had worked closely with the fishermen and taken their views into consideration but that "the fishermen did not agree with this proposal."[18]

Then Pete Haaker described how he'd witnessed the disease spread rapidly through the islands and to the mainland—knocking out 98 percent of animals at several locations. "The only hope for recovery," he explained, was "to protect the remaining disease-resistant black abalone."[19]

After the commercial divers voiced their views, one fish and game commissioner asked what would need to happen before the department reopened the fishery. The politic Petrovich explained that any commission closure would automatically sunset in just two years, but Haaker soberly added that, because black abalones were such slow-growing animals, "the fishery may not be able to recover for a number of years."[20]

The same commissioner asked Haaker point-blank if the number of afflicted black abalone had truly reached an "emergency level." Pete replied that the number of dying abalone had reached "a very high level" and, indeed, "constituted an emergency."[21]

In the end, the commissioners voted unanimously to close the state's black abalone fishery.[22] Haaker was relieved, but years later, he'd reflect with frustration, "It should have been done a lot earlier."[23] The precipitous decline of the blacks would become a somber prelude to other species' declines and an increasingly acrimonious debate over closing California's remaining wild abalone fisheries.

DEEPER TROUBLE

While Gary, Dan, and Pete had tracked the decline of sick black abalone in the intertidal zone, they'd worried about how the deeper-dwelling abalone species—reds, pinks, greens, and whites—were faring. Would withering syndrome strike them down, too?

With black abs, the biologists had witnessed the epidemic firsthand, watching as the familiar, living intertidal landscape changed dramatically. But gauging the status of the deeper-water species was more difficult. Monitoring Channel Islands National Park's kelp forests was akin to research conducted by state and university biologists at Johnson's Lee; researchers anchored their vessel, dove with scuba gear to a particular spot and depth, and then set and followed transects, noting on a small waterproof "slate" all the species they found within six to twenty meters of the transect lines to garner their data, a standard protocol for tracking changes through time.

By the early 1990s, the national park's dive surveys were showing fewer pink abalone in certain age classes, and CDFG surveys of San Clemente and Catalina Islands also found fewer pinks and greens. Four pinks at San Clemente Island showed signs of withering syndrome—some of the first evidence that the mysterious disease could afflict other abalone species.[24]

But they soon realized they had an even bigger problem—one species seemed to be missing entirely. Dan recalled someone saying, "I can't even *remember* when we last saw a white abalone."[25] They started to ask sport and commercial divers back at the marina and got a lot of shrugs. Some jabbed Pete with the obvious question: "Why is the fishery still open if there are so few left?"[26]

No one, it seemed, had seen a white abalone in years.

Japanese American diver Roy Hattori had first discovered white abalone at a depth of sixty feet near Point Conception in 1939, and veteran commercial diver Buzz Owens recalled that most white abalone he had harvested at Santa Cruz and Anacapa Islands in the late 1950s and early 1960s were taken at depths of less than eighty feet. But by the early 1990s, when the NPS-CDFG monitoring team started searching, whites were generally considered to be the deepest-dwelling abalone.[27] It would later become clear that the very deepest waters—130 to 200 feet—were the only places where white abalone had managed to survive.

Such deep waters presented a special challenge for surveys because at the time most of the agency's scientific divers were not certified to descend that far. But Gary and Pete were gung-ho divers—having logged thousands of dives each—and, it so happened, they were the official dive-safety officers for their respective agencies. With money already allotted for safety education, they decided to combine a series of depth-certification trainings with a survey for white abalone.[28]

Although scuba gear allows human divers to stay underwater by breathing a mix of oxygen and nitrogen, high levels of nitrogen can cause disorientation. This disorientation is amplified when divers descend below one hundred feet, an experience that has been called "the 'martini effect'" because it can feel like drinking a martini or two on an empty stomach. As Gary explained to me, "You *think* you can function fine, but, really, you can't."[29] In a worst-case scenario, an impaired diver could make mistakes that put his or her life at risk.

As such, deep-sea divers need to cultivate intense focus and pay careful attention to time spent underwater. At a sixty-foot depth, a scientific diver could stay underwater for sixty minutes, but at the safety limit of 130 feet, there was only fifteen minutes to descend, look around, and return safely to the surface, which means only eight minutes of bottom time. It wasn't efficient, but it was the only way to start surveying for white abalone.

During the summers of 1992 and 1993, the NPS and CDFG dive teams went out looking for white abs. They dove at fifteen locations within Channel Islands National Park waters where white abs had been previously surveyed and commercially harvested. All told, they searched thirty thousand square meters of suitable reef habitat. In an area that had reportedly harbored tens of thousands of white abalones in the 1970s, they found three.[30]

White abalone had been out of sight and out of mind. Now the animals were nowhere to be found.

✳ ✳ ✳

When Gary and Pete began to voice concern about the dearth of white abalone, many of their colleagues remained incredulous. As Gary recalled, "At the time, there was a pervasive belief that any animal that was a broadcast spawner—with such robust fecundity—could *never* be depleted by a fishery."[31] It was the same belief that California Bureau of Commercial Fisheries head N. B. Scofield had promoted so vigorously seventy years earlier—that there should be little reason for concern, since abalone produced millions of gametes each time they spawned. The tenacious idea persisted among fishermen and fisheries managers. Colleagues assured them: more white abalone must be out there somewhere—nestled on deeper, different, or more isolated reefs—though some did wonder if the mysterious withering disease that had zapped the black abalone could somehow be complicit.

When Pete recommended to his Department of Fish and Game supervisors that the state close the white abalone fishery immediately, he was again told that he'd need to conduct a full public process and environmental review. Some commercial divers, who used special gas mixtures to dive deep, claimed that white abalone persisted at depths deeper than the state biologists could reach with standard scuba techniques. These reports muted alarm and deterred department administrators in Sacramento from considering emergency action.[32]

Yet the field biologists knew that something was terribly wrong. Gary realized the only way they could figure it out was to go deeper, too. To do that, they'd need a submarine.

11.1. Concerned about the dwindling white abalone, biologists with the National Park Service and California Department of Fish and Game conducted surveys of deepwater reefs by submarine. Here, state biologist Pete Haaker slides into the hatch of the Delta in 1996. (Photo by Dan Richards)

* * *

It was, in fact, a yellow submarine—the *Delta*, owned by a company in Ventura that built and operated mini-subs for underwater research and filming. On behalf of their respective agencies, Gary and Pete submitted a joint research proposal and managed to secure a four-day slot to use the sub at the end of the 1996 season.

As the November stint approached, Gary planned how best to use their limited submarine time and lined up qualified scientific divers to round out a crew. When their launch date came, the divers boarded the *Delta*'s mother ship and headed out to deep waters where white abalone had last been seen— around Anacapa, Santa Barbara, and Santa Rosa Islands, and a distant submerged reef, Osborn Bank.[33]

The sub was small—only fifteen feet long and three feet wide—with a big, crab-pot-like bulge welded on top, and all manner of lights and cameras attached to its otherwise sleek exterior. Inside, there was room for only the company's pilot and one observer.

Pete Haaker remembered his first impression: "I'm going 200 feet deep in this?!"[34] To enter the cramped space, he had to slide down the hatch and then maneuver his legs behind him, lying flat in the belly of the craft with his face close to the front portholes. Then the pilot climbed in, sitting upright and straddling Pete's midriff, keeping watch from a porthole in the pot-shaped compartment on top. When the sub's hatch slammed shut, Pete felt claustrophobic, but when the vessel dove deep and then began to cruise along, he felt like he was flying.[35]

With the aid of lights, Pete could see several square meters of reef at a time—much larger than the view through a dive mask. And no wetsuit was needed; Pete wore a T-shirt and shorts. The submarine kept in close touch with the mother ship by radio. Pete kept his eyes focused on the passing reef-scape and described what he saw into an audio recorder; meanwhile, video cameras captured footage for follow-up, verification, and documentation.[36]

The biologists worked from dawn to dusk, cruising the edges of deep underwater reefs for miles. They took two-hour turns in the sub while those on deck reviewed video footage to determine how much area they'd covered and to spot any abalone that might have escaped notice. For the seasoned dive team, the world below 130 feet was new. Deeper than most kelp, the ocean seemed dim, quiet, and less lively, though an occasional ocean sunfish and even a California sea lion swam through their watery view.[37]

After a couple of days of searching, the team was excited when they finally found a single, live white abalone. The sub slowed and circled back to get its precise GPS location and film a good view, which was instantly relayed up to the rest of the crew onboard.[38] They hoped to locate some neighbors nearby but did not.[39] The research team ultimately found only four more white abalone.[40] Another submersible cruise the following October turned up only four additional animals.[41] After surveying nearly seventy-seven thousand square meters of habitat, the submarine surveys confirmed the biologists' greatest worry. Even in deep waters, very few white abalone remained.

Most important, each white abalone was solitary. One video clip from their survey shows the poignancy of one lone animal sitting on a singular pillar of rock jutting up in the midst of a vast empty space at a depth of 150 feet.[42] No one knew how far a white abalone's gametes could viably drift, but with no possible mate anywhere in sight, it was evident that this isolated creature would never get the chance to reproduce.

Moreover, all the white abalones that the sub team found were old animals—probably nearing thirty years in age. There was no younger generation poised to replace them. The sub team also found about three hundred empty shells, confirming that they were looking in the right places. Because all the animals (and shells) were roughly the same size, the biologists surmised that they'd all recruited around the same time, likely during a period with favorable ocean conditions back in the late 1960s or early 1970s.[43]

On their last evening, the crew kicked back on the ship's deck and considered what they'd seen underwater. Scientists generally take comfort in their data. A swarm of numbers can cast reality with the reassurance of statistical abstraction. But here, they'd come up with single digits. There was no comfort in a population size of four, or even nine.

Out there afloat on the dark water, with the pink glow of humanity hanging over Santa Barbara, Gary remembers wondering, Was their modest little mollusk a buffalo of the twentieth century or an underwater condor? Could the white abalone *already* be almost gone?[44]

The biologists had plunged to depths that few people ever plumb and come to face the grim possibility that they were seeing some of the last living white abalones on the planet.

* * *

Back at their desks in Ventura and Long Beach, the biologists pulled together all the information they had found about southern California's white abalone. Gary and Dan analyzed surveys done by a Scripps researcher in the 1970s that showed white abalones at patchy but high densities in fished areas.[45] Pete reviewed the CDFG's historic landings data and determined that whites had been first targeted by the commercial fishery in 1965, after pinks and reds started to decline. Landings peaked at 144,000 pounds in 1972 and then continued with roughly 44,000 pounds taken annually until 1976.[46] Then, white abalone dropped off the charts, literally. In 1978, the Department of Fish and Game eliminated the column for white abalone from its landings data with no explanation, presumably because there were too few to record.[47] That galled Gary. "The agency was supposed to be managing these resources for the public," he later told me, "so you'd think low numbers would have triggered concern. Instead, they just quit collecting data!" (This egregious mistake occurred during the department's late-1970s reorganization and budget cuts.) By Pete's analysis, 95 percent of all white abalones ever landed were harvested in just ten years.[48] Finally, the biologists crunched numbers from their surveys and confirmed the critically low density of the remaining white abalone they'd seen.[49] The few white abalones that the sub crew had located were miles apart.

Because early accounts suggested that white abalone might have succumbed to withering syndrome and some sick individuals were found, the biologists consulted with disease experts, who surmised that there would have been large numbers of empty shells in all different sizes had the animals died suddenly from disease. Instead, the pathologists concluded that, if the disease had killed off some white abalone, it would have been the final blow rather than the primary cause of the species' demise. The same-sized empty shells pointed to death at old age.[50]

Looking at the evidence arrayed before them, Gary, Dan, and Pete hypothesized that most white abalone had been unwittingly fished out back in the 1970s—leaving behind only a skeletal population of small, sublegal animals in the deepest waters. These isolated mollusks clung tight but were too far apart to ever reproduce. The biologists estimated that the entire remaining population of white abalone could be fewer than a thousand animals, with most approaching age thirty. Since white abalones had a lifespan of about thirty-five years, the clock was ticking for these aging individuals—and for the survival of the species.[51]

Feeling a weight of urgency, the agency biologists rushed to publish their ominous research in the journal of the American Fisheries Society. Their findings countered the long-standing but erroneous belief that marine invertebrates were resistant to overfishing.[52] Most important, white abalone were headed toward extinction unless something was done—and fast. "If we lose this species," they wrote, "we cannot claim we were not warned."[53]

Meanwhile, at the Department of Fish and Game, Pete Haaker and up-and-coming abalone biologist Ian Taniguchi assembled the data, conducted public outreach meetings, and prepared a formal environmental review. Though the situation with white abalone was most immediately alarming, Haaker recognized that the state agency also needed to take stock of what was happening to green and pink abalones and proposed a temporary fishery closure for those southern California species, too. "The number of animals removed by fishing is not adequately being replaced in kind by the production and survival of new animals," the department's environmental review warned. "Stocks have fallen to such low numbers as to endanger their continued existence."[54]

At a series of Fish and Game Commission meetings in the fall of 1995, about twelve commercial divers opposed the closure, some urging additional study and offering to help collect data. While some divers concurred with closing the severely diminished green and white abalone fisheries, they all opposed closing harvest for pinks, which they still could find in profitable quantities in one place: San Clemente Island.[55] However, with the state biologists' foreboding, deepwater research, and environmental review, the commissioners were growing increasingly concerned about the future of California's abalone. In December 1995, they voted to close the white, pink, and green abalone fisheries for two years—the limit of what the commission's authority allowed—to allow for more research.[56]

Everyone realized that an unfortunate side effect of closing fisheries for the most vulnerable blacks, whites, greens, and pinks would be to shift all fishing effort squarely to the most hardy and abundant abalone species remaining—the reds. They, too, had experienced erratic recruitment and shown signs of disease (albeit with less lethal effect) in southern California but had managed to persist in places with cold, upwelling water. Now these would be the only abalone left to take—for commercial and recreational divers in central and southern California, and for sport divers on the North Coast.

The biologists feared that reds would be the next species to crash in what was beginning to look like a grand abalone train wreck. It was only a matter of time before they'd need to rein in that fishery, too. Only this time, the push for action would come from entirely different quarters.

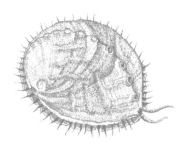

CHAPTER 12
The Abalone War

When Paul Turnbull came to the front of the room, the members and staff of the California Fish and Game Commission braced themselves. The sport diver and abalone advocate had become a regular at the commission's meetings, warning again and again that the commercial fishery was decimating the state's red abalone.

A draftsman from Hayward, Turnbull had been sport diving for reds on the Central and North Coasts for thirty years, and even had a mammoth 11⅜ inch trophy abalone to his name. But since the state had opened the San Mateo coast, south of San Francisco, to commercial divers in 1990, he'd witnessed his favorite dive spots picked clean of large abalone. He volunteered for a while as conservation director for the Central California Council of Dive Clubs before starting his own group, the Abalone and Marine Resources Council (AMRC). The small, motley crew of East Bay divers met at a pizza shop to exchange ideas about how to stop the hastening decline of the shellfish they loved. Soon, they garnered six hundred supporting members.[1] With his bellowing voice, Turnbull aimed to represent them all.

His prime tack at the meetings was to ask pointed questions. "Why haven't the state wardens been provided more support to stop illegal commercial poaching operations in Sonoma County? Isn't it true that some commercial divers have parked their boats just outside the boundaries of the Fitzgerald Marine Reserve, then illegally swam underwater into the protected area to poach abalone?" Persistent and abrasive, he dug in: "Why are commercial divers with air assist allowed to take 84 abalone per day per person . . . when sport divers using snorkels are only allowed to take four? Why should the Commission wait for the abalone . . . to be wiped out before taking an action?"[2]

No one ever answered him. According to *San Francisco Chronicle* reporter Tom Stienstra, Turnbull's "endless one note samba warning that commercial diving was wiping out the abalone had turned him into a pariah."[3]

The concluding thrust of Turnbull's impassioned appeals was a call for the commission to put an end to commercial abalone diving until the California Department of Fish and Game (CDFG) developed a scientifically based fishery management plan for abalone. He cited state code that gave the commission authority to safeguard resources from depletion. No matter, the legal counsel would advise that the fish and game commissioners didn't have authority to close only the commercial abalone fishery without also closing the sportfishery. The implication was clear: the politics of taking action was impossible.[4] The commissioners would punt and proceed to their next agenda item.

However, in the fall of 1996, Turnbull upped the ante. He formally demanded that the commission make an emergency closure of the commercial abalone fishery until the state developed a fishery management plan. And this time, he had backing from recreational dive clubs to shut sportfisheries south of San Francisco, too. As one northern California dive club leader would later recall, it was "like Ducks Unlimited asking to stop duck hunting."[5]

The commission could no longer dismiss the public call for abalone conservation. The shift was tectonic, but the underlying tensions and pressures had been building for a long, long time.

* * *

The year before, in 1995, Rocky Daniels was out diving. A firmware engineer from Santa Rosa, Daniels spent many of his weekends encased in a thick, black, neoprene wetsuit. Slipping into the clear, cold water with friends, he pressed through lush kelp to explore the underwater caves and reefs of the Sonoma and Mendocino coastlines, often hunting abalone.

This time he was on a dive club spearfishing trip. Club members launched inflatable boats from the beach and headed to a reef about a third of a mile offshore, known for its unusual underwater topography, with jumbles of car- to house-sized boulders. As it turned out, visibility was terrible—Daniels couldn't even see his neon-green dive fins—so he gave up on spearfishing and switched to exploring the pockets between the big rocks with his dive light. Some pockets were "chockablock with large abalone," but in several others, he found only bare, oval shaped scars—spots where the slow-growing but now-missing animals had been clinging for years. Because he was so far offshore, "it was crystal clear that I'd hit on evidence of commercial-scale poaching," Rocky later recalled, "and it seriously pissed me off."[6]

Poaching had long been an issue on California's North Coast, but the problem had intensified as abalone became scarcer in the south. With each abalone worth up to $50 on the black market, poachers were willing to take greater risks. Increasingly they targeted northern California's coast, where red abs remained abundant owing in large part to the long-standing policy that excluded commercial harvest and allowed only breath-hold sport diving—leaving a deepwater refuge beyond the reach of free divers where adult abalone could grow, reproduce, and continually replenish the more accessible stocks.[7] (Though in the late 1950s CDFG biologist Keith Cox's surveys had found offshore red abalone did not move inshore, subsequent and more substantial tagging studies conducted over a seven-year period in the early 1970s by Humboldt State marine biologists John DeMartini and Jerald Ault had shown that red abalone did move inshore during winter months, likely following food as bull kelps ripped out by storms washed ashore.[8])

Each year, state game wardens made dozens of arrests, nabbing poachers who filled car trunks with several "limits" and sold them to markets and Chinese restaurants throughout the Bay Area through illicit, difficult-to-track channels.[9] But the fines were small compared with the money that could be made, and for every bust, many more poachers went undetected. In response, local sport diver George Lawry started the Sonoma County Abalone Network (SCAN) to help the chronically understaffed fish and game wardens watch for poachers and follow cases through the court system. Rocky became one of the group's key volunteers. SCAN carried forth the North Coast's deep-rooted tradition of protecting its local abalone.[10]

∗ ∗ ∗

Not long after Rocky discovered the ransacked offshore reef, CDFG warden Dave Bezzone made a big bust. Working undercover, often on his own time after-hours, Bezzone had watched at Timber Cove, north of Fort Ross, and pieced together evidence for nearly two years. Eventually, he tracked twelve people from Santa Rosa who had been illegally using scuba gear to dive for abalone, shucking them underwater, stashing them in burlap bags, whisking them by van to the San Francisco airport, and then flying them to Los Angeles, where the ringleader, twenty-seven-year-old commercial fisherman and seafood wholesaler Van Howard Johnson, supplied southern California restaurants and shipped the rest to New York, Japan, and China.

Late one night, after listening with an electronic bug to logistics planned for the next day, Bezzone arranged for Johnson to be apprehended at the Los Angeles airport with a big shipment of illegal abs.[11] Ultimately, the investigation determined that this one ring had illicitly taken more than thirteen thousand abalones from Sonoma's reefs in less than a year. According to the Department of Fish and Game, it was equivalent to 18 percent of the county's annual, legal, sport take—a damaging chomp into the North Coast's abalone stocks.[12] The Timber Cove bust turned out to be the largest ever.

Though poaching was typically regarded as a misdemeanor, with only minor fines, Sonoma County district attorney Brooke Halsey prepared to prosecute the egregious case of multiple violations as a felony conspiracy to ensure stiffer fines and punishments, aiming to deter future poaching. Northern California sport divers followed the case closely. When the judge announced a plea bargain deal with the poachers that gave ringleader Van Howard Johnson only ninety days in jail and a $40,000 fine, they became incensed. Many considered the fine—just a small percent of the harvest's worth—scarcely "a wrist slap," sure to encourage more poaching.[13] Daniels enlisted SCAN members and started a broader letter-writing campaign to the judge, pressing for stiffer sentences. Forty angry divers from all over northern California showed up at the sentencing hearing to protest, as did community members, a local television camera crew, and newspaper reporters.

In response to mounting public pressure, the judge withdrew the plea bargain deal and instead sent the poachers to prison. Two, including ringleader Johnson, withdrew their guilty pleas and let the cases go to trial. Ultimately, after a four-week trial, the judge sentenced Johnson to three years in prison and fined him $50,000. It turned out to be the largest prosecution of an abalone poaching case ever.[14]

Daniels was not the only one who'd seen devastation wrought by poachers. Many North Coast divers grew up in families that came to the Mendocino and Sonoma coasts to hunt abalone every year—for generations. For many, it was "almost a sacred ritual."[15] At the end of spirited days, they peeled off wetsuits, camped in redwoods, pounded abalone on stumps, shared stories, and savored unpretentious seafood feasts. They felt protective of the North Coast abalone stocks that anchored their tradition, and they wanted to do something more. The Timber Cove case outraged sport divers and united them around the goal of conserving abalone in a whole new way.

* * *

Years earlier, in 1990, another notable bust had kindled the sport divers' smoldering anger: several commercial urchin divers were convicted for poaching abalone. After the 1982–1983 El Niño in southern California, commercial divers, many of whom carried permits for both abalone and urchin diving, persuaded the Fish and Game Commission to open the North Coast to red urchin harvest.[16] Shortly thereafter, while analyzing landing data, Fort Bragg–based Department of Fish and Game biologist Konstantin Karpov noticed a suspicious increase in the amount of abalone that commercial divers claimed they were landing in southern California. In 1983, the largest commercial landings of abalone reported by single boats were twenty-one dozen in the San Francisco Bay region and twenty-two dozen in Southern California. In 1987, the largest landing near San Francisco was again twenty-one dozen, but the largest south state landing had jumped to forty-seven dozen. Since landings everywhere had been dropping for years and department biologists' surveys had just documented severely depressed abalone growth in the wake of southern California's El Niño, the marked uptick was a giant red flag.[17]

Karpov suspected that some unscrupulous commercial divers were illegally loading up on abalone while scuba diving for red urchins on the North Coast, then cruising by way of the Farallon Islands, where commercial abalone harvest was legal—before landing their catch in Half Moon Bay. Or the divers might transfer North Coast abalone at sea to boats that illegally shipped and landed them in southern California ports.[18] Karpov alerted local wardens, and, sure enough, they soon busted two commercial urchin boats for poaching North Coast abalone. The commercial divers, including one who had become somewhat of an industry spokesman, had converted scuba tanks into secret abalone storage containers.[19]

In 1990, the legislature responded to public outrage by sharply increasing penalties for poaching, yet after a short hiatus, Karpov found another jump in southern California abalone landings. Northern California's red abalone were still flowing illicitly into the commercial marketplace.[20] Meanwhile, commercial divers—facing depleted beds in the south—had continued to press to open an experimental commercial abalone fishery in northern California in the vicinity of Point Arena. Local businesses, including Fort Bragg's Sub-Surface Progression dive shop, collected thousands of signatures on a petition in opposition. Responding to his constituents' concerns, when Assemblyman Dan Hauser worked on the 1990 abalone bill, he added language that finally outright prohibited commercial abalone diving on the North Coast. But the compromise bill also ended up opening a new area for commercial take—the reefs of the San Mateo County coast—putting commercial divers into direct competition with Bay Area, Central Valley, and local sport divers, who had long favored the area for recreation.[21]

Finding fewer abalones, sport divers there soon became incensed that commercial divers, who could use surface-air-supply breathing gear (known as hookah), were decimating the reefs. In 1993, when Karpov and other Department of Fish and Game biologists conducted dive surveys of Fitzgerald Marine Reserve near Half Moon Bay, they found that abalone "densities were remarkably low compared to other northern California areas."[22]

SCAN and the Abalone and Marine Resources Council began to engage more North and Central Coast sport divers. They reported on poaching incidents and Department of Fish and Game research in their newsletters and hosted talks by Karpov, who had become convinced that poaching was putting northern California's prized abalone stocks at grave risk.[23] The open abalone fisheries at the Farallon Islands and along the San Mateo coast not only made it possible for commercial divers to land illegal quarry from northern California at legal ports, but also made it harder for wardens to bust poachers. Moreover, as abalone became increasingly scarce, Karpov found, its price continued to climb, boosting the incentive for poaching. In an internal CDFG report detailing the problem, Karpov wrote, "Illegal harvest in Northern California at levels implied by our review cannot continue at the expense of sportfishery resources estimated to be worth $11.5 million dollars annually between Bodega Bay and Cape Mendocino."[24]

For decades, commercial abalone divers had tried through political channels to break into the rich reefs of the North Coast to no avail. Now it was clear that some of them, at least, were breaking in through illicit means. Moreover, poached North Coast abalones could too readily become "legal" and salable in the voracious global marketplace. To sport divers, it was looking more and more like commercial interests were closing in on northern California's cherished abalone grounds—the state's last bastion of abalone abundance. No wonder that Turnbull and his East Bay diving buddies and then Daniels and his SCAN friends had become outraged.

However, the critical mass for change would ultimately come with new allies from the south.

* * *

Growing up in southern California in the 1960s, Steve Benavides spent big chunks of every summer on Catalina Island. His family had a boat, and twenty-five weekends a year, ten kids plus friends piled on and cruised out to Catalina. While his sisters sunbathed and played the radio on deck, Steve and his big brother escaped to the cool, quiet underwater world. His brother had built a scuba tank with two regulators so they could go down together. They spent hours exploring underwater near Avalon and Cherry Cove Canyon. On the north side of the island, big submerged boulders were covered with green abalone. Steve fell in love with diving and developed a deep connection to Catalina. But over time, he watched the island's abundant fish, spiny lobster, and abalone dwindle. Like many recreational divers in southern California, he was compelled to switch from abalone hunting and spearfishing to photography. Eventually, he started to write about his underwater adventures in *California Diving News*, as a countercurrent to his day job as a CPA and tax attorney.

Benavides got his first taste of the acrid politics of marine fisheries when curiosity drew him to attend a Fish and Game Commission meeting in Long Beach in the late 1980s. He quickly ascertained that "commercial interests ran the show," and that put the decline of fish and abalone he'd personally witnessed into a disturbing new perspective.[25]

Soon thereafter, he became involved with a statewide ballot initiative to end gill-net fishing, a commercial practice that captured sharks, dolphins, and other sea life as bycatch along with target fish.[26] He started by giving slide-shows to dive and nature clubs in the Los Angeles area. Then, he became a founding member of the Catalina Conservancy, a group dedicated to creating marine parks around the island. He also helped a friend to end aquarium-trade fishing, a practice that scooped up thousands of colorful gobies, garibaldis, leopard sharks, and other beautiful fishes from the island's nursery reefs. With some conservation successes under his belt and a growing appetite for political activism, Benavides was ready to take on something bigger.

He'd first free dived for abalone in northern California when he attended college in San Francisco, and periodically returned. On the North Coast, he saw what he'd remembered from his childhood on Catalina—undersea reefs studded with ample abalone. On trips north in the mid-1990s, he got to know some local sport divers. They explained the poaching problem—how the commercial market in southern California, with its direct channels to high-value Asian markets, created a huge incentive for illegal take of northern California's still abundant abalones. They described how commercial divers had tried to break into reefs near Point Arena and had elbowed their way into the San Mateo coast's Fitzgerald Marine Reserve. At the time, Benavides recalled, they all believed that the commercial industry was the biggest problem for abalone. It became clear that North and South Coast sport divers held many frustrations and concerns in common.

Steve started to talk on the telephone with dive club leaders, including Paul Turnbull and Gene Kramer of AMRC and Steve Campi, president of the Central California Council of Dive Clubs (Cen-Cal), one of the state's largest marine fishing and hunting groups. Taking a statewide perspective, the dive leaders believed that the key to conserving California's remaining red abalone—on both the North and South Coasts—was to turn off the relentless demand of the commercial market by closing the commercial fishery entirely.

Through Turnbull's fruitless efforts at the Fish and Game Commission, they already knew they couldn't propose to close the commercial fishery south of San Francisco without also closing the recreational fishery. Most were ready to take that step. Though Turnbull wanted to keep Fitzgerald Marine Reserve open to sport divers, all the other dive club leaders recognized the powerful message they could send by supporting the closure of their very own fishery. And with so few abalone left for recreational divers in central and southern California, there was little to lose and much to gain: the possibility that kids or grandkids might someday have the chance to forage for abalone in southern California again, and better management of northern California's abalone to ensure a fishery for sport divers there into the future.[27]

Turnbull's plan to push for an emergency fishery closure at the Fish and Game Commission's Eureka meeting galvanized the dive club leaders' support.

✳ ✳ ✳

The *San Francisco Chronicle* predicted it could be "the biggest showdown of the year over natural resources in California."[28] It would also be a landmark case because it was the first time a user group demanded emergency take restrictions on its own quarry, and the first time a sport group demanded a management plan for a marine species.[29]

In his request on behalf of the Abalone and Marine Resources Council, Turnbull underscored the urgent need to limit diving permits "to protect the rapidly diminishing red abalone resource. . . . This kind of action is necessary to protect both commercial divers and sport divers alike."[30]

When commissioners discussed the AMRC's request, Commissioner Richard Thieriot expressed concern about the growing public perception that the commission was always behind the curve, acting to protect natural resources only after the "damage has already occurred." Turnbull again pressed his case that the Fish and Game Commission was bound by law to direct the department to develop a fishery management plan for sustainable harvest. "We're saying here and now it's time to close commercial diving to keep them from wiping out the abalone."[31]

When the commission's legal counsel once again explained that the commissioners couldn't close the commercial fishery without also closing the sportfishery, Turnbull goaded them to go ahead and do it. This time he had backing from California's biggest sportfishing group and dive clubs throughout the state.[32] Representing southern California's Catalina Conservancy Divers, Steve Benavides had sent testimony asking the commission to close the entire coast south of San Francisco to all abalone take. "The landings of abalone have plummeted in the past 10 years and the fishery is not operating on a basis of sustained take," he wrote. "The Department of Fish and Game's own figures show the resource to be in a perilous drop."[33] Larry Ward of United Anglers Conservation Council also (now United Anglers) provided testimony, contending that the state's lack of fishery management plans had allowed a small number of commercial fishers to dominate and devastate several species, including abalone. "The rights of California's citizens to enjoy the wide diversity of angling opportunities must be preserved and must not be threatened by destructive commercial interests," Ward argued.[34]

Without the usual "out," the Fish and Game Commission and the department's Sacramento political staff were backed into a corner. The CDFG's deputy director for policy, Al Petrovich, explained that his agency didn't have sufficient funding or personnel to develop a fishery management plan for red abalone. Instead, he suggested that the existing ad hoc abalone stakeholder committee consider the AMRC's proposal more carefully and report back with a better-informed recommendation in February. Some regarded it as a ploy to dodge the issue, but the commissioners had recognized the significance of the broad constituency of conservation-oriented sportfishing and dive clubs asking for their own fishery to be closed.[35]

Just a few weeks later, right before Christmas 1996, the *Sacramento Bee* published a landmark series that would profoundly change public perceptions about abalone and ocean issues in California. In "Pacific Blues," Pulitzer-prize-winning journalist Tom Knudson and Nancy Vogel depicted the state's ocean as utterly beleaguered. With broad strokes, their articles identified a range of horrific problems. Overfishing topped the list. California's fisheries had declined by 70 percent in the sixty years since the 1930s, with rockfish on the verge of collapse. The threat of poaching earned its own front-page article, with abalone as the poster child. Knudson characterized the labyrinthine poaching cases with their cunning James Bond–like plots: "There are midnight landings, safe houses, false-bottom containers, fraudulent permits, fictitious names and lookouts with hand-held radios." All in all, he explained, two million abalones were taken each year in "a huge seafood heist." Knudson and Vogel concluded, "Wardens are winning some battles. But government is losing the war."[36]

As devastating as overfishing, pollution, and poaching were, the most troubling matter of all was that no one in government was addressing any of these problems in a meaningful way. The article singled out the state's Department of Fish and Game for its record of mismanagement but also pointed to the fact that the legislature had slashed the agency's budget to the bone, cutting a whopping 60 percent in the six years since 1990—and that after even earlier shavings. With dwindling funds for fish and wildlife, there were too few wardens to enforce conservation rules. Moreover, state biologists' recommendations had routinely gone unheeded by the Fish and Game Commission, and no one had a grip on what was happening. The legislature—charged with managing most commercial fisheries—also had a dismal record. One fishery after another had collapsed as assembly members narrowly guarded the short-term interests of their own district's fishers and never looked at the big picture. Knudson and Vogel pulled no punches, warning that Californians were "squandering their coastal heritage." The *Bee* editorialized that the state was penny-wise and pound-foolish to ignore the health, value, and importance of its unique ocean resources.[37]

"Pacific Blues" struck a chord. Many individuals recognized that their own firsthand experiences of loss—Turnbull's picked-over reefs, Rocky's cleared-out pockets, Steve's bare walls at Cherry Cove Canyon—were part of a troubling larger trend. It was time to jump on the bandwagon and push for change. And, finally, leaders in state government wanted to be on the side of conservation.

*   *   *

With Turnbull's opening and the publication of "Pacific Blues," the dive club leaders pulled together. Public use of the internet had just started, and so SCAN's Rocky Daniels set up a private network for the dive leaders to communicate. Over the course of several weeks, they discussed issues, developed strategies, and planned a new four-point campaign for abalone. They'd press for an emergency closure of all abalone fishing south of San Francisco; push for a new abalone stamp on state sportfishing licenses to raise money for enforcement, research, and restoration; advocate to further increase penalties for poachers; and resist—at all costs—any new attempt to open the North Coast to commercial abalone fishing.[38]

To make their case for the fishery closure, the dive club leaders knew they'd need strong scientific backing. Through his work with the Catalina Conservancy, Benavides had come to know southern California abalone biologists Mia Tegner from Scripps and Pete Haaker from CDFG. He became familiar with their research and the new science that emerged after the international symposium underscoring abalone's vulnerability. Daniels knew CDFG biologist Konstantin Karpov from working with SCAN in northern California. The sport divers tapped the biologists' expertise, studied up on the latest findings, and began to assemble the science needed to support their proposal.

The dive club leaders also knew they'd need strong grassroots support, so Rocky expanded their electronic communication network beyond the core group to include more dive club leaders and then dive club members throughout the state. Recreational divers were just beginning to use the internet to organize outings, so Rocky was able to garner their interest. His email list grew from three to nine to sixteen and, eventually, to twenty thousand. Ultimately, he would send out alerts motivating sport divers to show up at crucial meetings and to write letters to their elected representatives before key hearings. The sport divers' abalone protection effort would become one of the first conservation campaigns to use the internet.[39]

The dive club leaders plotted two fronts. First, they would press the Fish and Game Commission to close all abalone diving south of the Golden Gate and at the Farallon Islands. At the same time, they would pursue a bill in the legislature to accomplish the same aim, plus increase penalties for poaching and start an abalone stamp program to fund research and more wardens. Benavides recalled that the core team called themselves the Able Group, based on a quote from Sun Tzu's *Art of War*: "He whose generals are able and not interfered with by the sovereign will succeed." Indeed, when hostilities intensified during their two-pronged campaign, the dive club compatriots would come to call the conflict the "abalone war." According to Benavides, they may not have realized the difficulty of their challenge at the outset, but they were completely convinced they would succeed.[40]

* * *

When the Fish and Game Commission reconvened in Monterey in February 1997, the Able Group was ready. Turnbull, Kramer, Benavides, and about fifteen other sport divers turned out to speak in support of closing the coastline south of San Francisco to all abalone fishing, aiming to conserve the state's red abalone before it was too late.

This time, commercial divers showed up in force, too, and twenty-two—more than one-fifth of the licensed abalone divers remaining in the state—spoke in defense of their industry and their livelihoods. They staunchly opposed the closure. From the commercial divers' perspective, there were still "plenty of abalone"—if you knew where to look. At the time, they were landing red abalone from three remaining strongholds: Point Conception, the Farallon Islands, and San Miguel Island—remote locales that still had relatively abundant stocks compared to everywhere else along California's 1,400-mile coastline.[41]

After weathering closures of the black, white, pink, and green abalone fisheries, John Colgate, president of the California Abalone Association, questioned the Department of Fish and Game's science and urged the commission to get a "second opinion." Commercial divers thought that the biologists' method of surveying random transects to track changes in abundance over time—the standard scientific protocol for marine monitoring—made no sense for abalone and accused biologists of spinning data against them. Colgate also explained that the abalone industry had contributed hundreds of thousands of dollars to outplant hatchery juveniles to sustain the fishery. Other commercial divers accused sport divers of having more poaching violations than the commercial divers, and some, once again, proposed opening the North Coast to commercial abalone harvest to ease fishing pressure in the south.[42]

Although most professional abalone divers, at that point, had shifted to fishing primarily for red urchins, those who focused solely on abalone would be hit hardest by a closure. One diver, who'd recently invested in the two permits needed to join the limited-entry abalone fishery—plus a boat and dive gear—pleaded that he'd lose everything he owned. A diver's wife from the Women's Auxiliary of the California Abalone Association implored the commissioners to make "a compromise not a closure" so as to prevent economic hardship for commercial divers' families.[43] The commission listened but wanted to wait for a clear recommendation from the ad hoc abalone stakeholder advisory committee before making any decision.

Meanwhile, the stakeholder group had made little headway. At its first meeting in San Jose, the department's biologists, Pete Haaker and Konstantin Karpov, collaborated on a presentation to committee members and a packed room of commercial and sport divers, explaining the many factors that had contributed to the red abalone's demise in central and southern California. They emphasized that it was an oversimplification to blame overharvest alone but underscored that, with fishing pressure now focused in just a few small areas, remaining red abalone faced a much higher risk. "Unless remaining stocks receive some form of protection," Karpov explained, "localized disappearance of red abalone in central and southern California is likely."[44] After commercial and recreational divers presented their views, a facilitator helped the group identify twenty-three options, but at the end of a long day, the ad hoc committee failed to reach consensus.[45]

At a second meeting in Long Beach, Haaker and Karpov described the grim results of their early February dive survey at San Miguel Island. Although they found evidence of good recruitment in some places, they'd also found several red abalones infected with withering syndrome—a worrisome sign. After assessing 3,200 abalones, they'd determined that only 1.2 percent of animals were large enough for commercial harvest.[46] Haaker called that "pretty low to support a commercial fishery."[47] The department biologists urged the stakeholders to support the proposed protective closure.

However, commercial divers remained adamant that San Miguel Island harbored enough abalone to supply their continued take. Sport divers countered that all other red abalone populations south of San Francisco had dropped so low that San Miguel's abs would be needed as a source of broodstock to restore other sites. Commercial divers insisted that sea otters, pollution, and poaching were to blame for abalone declines, but sport divers maintained that commercial take was the biggest remaining threat—and the only one that could be readily addressed. Ultimately, the only consensus the ad hoc committee could reach was to recommend that the Department of Fish and Game complete a full environmental study of all abalone species as a way to make sure that science would guide management.[48]

When the Fish and Game Commission reconvened in March in Sacramento, expectations were high that the commissioners would take some definitive action on the closure. After listening to hours of impassioned testimony from both sides, they decided to follow the ad hoc committee's recommendation and directed the department to complete a comprehensive environmental review by the end of the year.[49]

However, Commissioner Richard Thieriot, who had presided over the eleventh-hour closure of all the other abalone species, was frustrated by his colleagues' foot-dragging. "I fear that when we say, 'Well, let's study some more,' we fall into the trap of moving at the last possible minute," he explained. "Since we do not move fast, there are no pink abalone for (commercial divers). There are no green abalone. (Commercial divers) are the ones who are going to feel the pain. . . . It's our job to husband the state's wildlife resources to prevent things from getting to this point."[50] Commissioner Frank Boren, who had also presided over the other abalone closures, concurred. "In my six years here, we have *never* closed in time," he reflected. "Somebody has to have the courage to do something that people don't like." Thieriot proposed to his fellow commissioners that they reconsider an emergency closure in May. All agreed, but they requested more substantive input from the department's biologists and scientists employed by the abalone industry.[51]

✳ ✳ ✳

Because the Fish and Game Commission could at most issue only a temporary, two-year closure, the dive club leaders had also started to work a second front in the legislature, where they found a willing sponsor in state senator Mike Thompson. A Democrat who represented Sonoma and Mendocino Counties, Thompson had grown up in a family that headed to the coast each spring for abalone. He knew firsthand the importance of the iconic shellfish to his constituents and recognized the ways the state had failed in abalone management. He understood, too, that the sport divers' grassroots organizing had created a unique opportunity for political action.

In early 1997, Thompson had already introduced a bill to again increase penalties for abalone poaching. When dive club and local business leaders encouraged him to add provisions similar to those being considered by the Fish and Game Commission, Thompson was ready to take the lead. Meanwhile, Republican governor Pete Wilson's administration sponsored a high-profile conference about the future of California's ocean resources while legislators—including many Democrats recently elected to majority with a "vote the coast" campaign sweep—introduced dozens of bills, riding a cresting wave of bipartisan support for coast and marine conservation.[52]

That was the backdrop in mid-April when the Senate Natural Resources Committee held its hearing on Thompson's abalone bill. This time, when sport and commercial divers headed to the capitol, the stakes were much higher: a legislative closure of abalone fishing south of San Francisco could be permanent, and whatever happened with this key committee would likely influence the Fish and Game Commission and also subsequent consideration in the full senate and assembly.

There is no transcript that captures what was said at the hearing, but those who attended recall that the day was suffocatingly hot and the room was charged with edgy tension. While everyone could agree on the need to increase fines for poachers, to develop a science-based fishery management plan, and to create an abalone stamp program to generate more funds for wardens and research, the proposal to close all abalone fishing south of San Francisco remained anathema to the commercial divers. The written testimony submitted to legislators, now held in the state archives, offers a window into the arguments aired.

The committee invited Scripps scientist Dr. Mia Tegner to testify as an abalone expert. With nearly three decades of marine research experience in southern California and fourteen scientific publications about abalone, she explained with unequivocal authority how red abalone populations—once plentiful off many of the Channel Islands—had perilously declined and how commercial divers had long maintained profitable abalone landings by serially expanding into new harvest areas. But with all the fishing pressure now focused on only a few areas, remaining populations were at greater risk. "The faster we close the fishery, the greater the chances of rehabilitation," she urged. "I fear if we wait too long, we will cross a threshold from which recovery is not possible."[53] Tegner never blamed the commercial fishermen for the red abalone's demise but emphasized that the onslaught of many different stressors had created a crisis that demanded urgent action. She admonished the legislators to apply the lesson of the nearly extinct white abalone to red abalone before it was too late.

The California Abalone Association (CAA) tapped not a marine biologist but UC Berkeley statistician Philip Stark as their science expert. Dr. Stark argued that abalone populations were not in trouble and that long-term declines in commercial landings had resulted primarily from smaller numbers

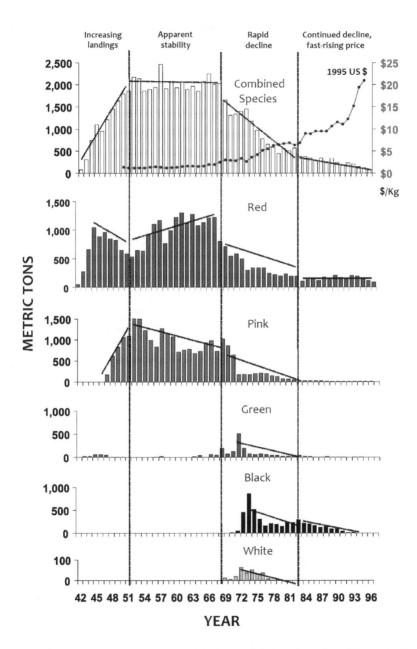

12.1. In the 1990s, CDFG scientists analyzed the historical decline of commercial abalone landings. These nested graphs show how the department's reliance on tracking overall landings obscured the serial depletion of individual abalone species. (after Konstantin A. Karpov et al., "Serial Depletion and the Collapse of the California Abalone (*Haliotis* spp.) Fishery," in Workshop on Rebuilding Abalone Stocks in British Columbia, 2000.)

of divers working shorter seasons. He also contended that the department's practice of conducting surveys at fixed locations did not produce an accurate accounting of actual abalone populations.[54] Stark's narrow line of reasoning stood in sharp contrast to Tegner's depth of experience.[55]

After the experts spoke, the usual suspects made their arguments. Leaders from dive clubs from San Diego to Mendocino, including Benavides, Daniels, and Campi, voiced concern about the future of the state's abalone, urged the closure, and pressed for even higher abalone stamp fees to ensure sufficient funding for wardens and research. In his testimony, Gene Kramer, who had taken over from Turnbull as head of the Abalone and Marine Resources Council, explained, "All we are asking for on this resource . . . is that the level of take be set so that the population remains somewhat stable or increases. . . . We could have had a sustainable commercial harvest of perhaps a million pounds per year, if the divers of 30, 40 and 50 years ago had set their harvests lower."[56]

On the other side, CAA president John Colgate urged the state senators to leave management of the commercial abalone fishery to the Fish and Game Commission. CAA secretary Michael Kitahara argued that the bill unfairly blamed commercial fishermen while doing nothing to stop poachers. He explained that recreational divers now took two million red abalones each year—far more than commercial divers—and underscored the economic hardship some fishermen would face. Other commercial divers self-identified as environmentalists, describing their fishery as "one of the cleanest" and explaining their efforts to reseed abalone through outplanting. All called for better management of abalone instead of a closure.

However, in the end, the senators of the Natural Resources Committee were moved most by the risk of losing abalone—already depleted, species by species, one area after another until almost nothing was left—and voted six to one for the closure bill, all but assuring it would sail through the full senate. Recreational divers believed Mia Tegner's powerful words carried the day.[57] While many scientists took pains to avoid the disagreeable crucible of political engagement, Tegner did not shirk. For the fate of abalone, that would make all the difference.

* * *

Three weeks later, the warring sport and commercial diver factions trooped back to Sacramento to make ever-more-impassioned pleas at the Fish and Game Commission's May meeting. With the Thompson bill advancing in the legislature and an emergency fishery closure on the agenda, the conflict intensified.

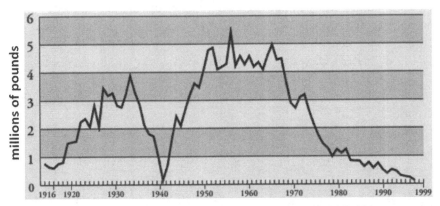

California Abalone Landings, 1916-1997

12.2. California Department of Fish and Game landings data show the rise and fall of the commercial abalone fishery from 1916 to 1997. As long as the catch continued to climb to its peak in 1957, managers presumed that size limits and closed seasons were working to sustain the fishery. In retrospect, we know this laissez-faire approach—common in fisheries management at the time—was deeply flawed, neglecting to account for differences in individual abalone species, the high variability of abalone growth and reproduction, place-specific effects of rising fishing pressures, and environmental stressors. (after CDFG, California's Living Marine Resources Status report, 2001)

This time, following the commissioner's request for more scientific input, the CDFG's biologists testified. Pete Haaker was more comfortable on a dive boat, but he explained the abalone biology he'd come to know so well. Although abalone could be prolific spawners, recruitment of new animals was often highly erratic owing to intense predation on juveniles, variable food supplies, and changing ocean conditions that affected larval survival. Plus, he explained, the long-lived, slow-growing, sedentary animals were highly vulnerable to overfishing. According to the department's data, red abalone landings had declined 87 percent since the fishery's heyday. Haaker underscored that the perilous condition of black, white, pink, and green abalone stocks had already necessitated closing those fisheries, now putting reds at higher risk.[58]

Haaker's North Coast counterpart, Konstantin Karpov, followed up, explaining that abalone needed to be clustered close together to reproduce, but recent surveys had found low densities at all central and southern California monitoring sites. Populations remained strong only in northern California, where commercial harvest and scuba had long been prohibited, protecting abalone with "refuge of depth" beyond the reach of free divers.[59] "Given the kind of decline we've seen," Karpov warned, "we're playing brinksmanship with the species."[60]

Next, the CDFG's shellfish disease expert, Dr. Carolyn Friedman, explained the devastation of abalone by withering syndrome—how warm water exacerbated its spread and afflicted all of California's abalone species in ways she was still investigating. Though reds were more resistant than blacks, the full effects of the pathogen remained unclear. She also discussed sabellid worms, nonnative shell-boring pests inadvertently introduced into California's abalone aquaculture facilities from South Africa, which had, in one recent case, escaped into the wild from an abalone farm's outfall pipes and infected an array of intertidal gastropods. The crisis had prompted the department to place a moratorium on all abalone outplanting, given the potentially high risk of spreading the damaging pest.[61]

Then Scripp's ecologist Mia Tegner again spoke as an expert scientist. "Throughout Southern California, island by island," she explained, "the (catches) have crashed where there once existed productive abalone beds." She told the commission about a recent survey in which biologists had found only fourteen red abalones during eight hours of diving. "After San Miguel is depleted," she warned, "there'll be no place left for the commercial fishermen to go."[62] Dr. Tegner also preemptively countered the divers' contentions that outplanting hatchery-reared juveniles could rehabilitate and sustain abalone stocks for the fishery.[63] Years of intensive research had convinced her otherwise.

Then those opposing the closure were invited to speak. The California Abalone Association's attorney, Ilson New, testified first. He argued that landing data were a poor indicator of red abalone's status and contended that the state's biologists were "biased" and had not followed proper methods in their research. He admonished that a closure would render the commercial divers' equipment useless, which could be considered "a form of illegal take by a governing body," and urged commissioners to base their decision "on science and facts and not emotion."[64] UC Berkeley statistician Dr. Philip Stark again served as the commercial divers' expert and again argued that declines in landings reflected fewer fishermen catching fewer abs in shorter seasons rather than declines in actual abalone populations.[65] Their testimony did not address the growing weight of evidence about abalone's vulnerabilities.

The sport and commercial factions once again reprised their arguments, as well. Sport divers pressed the commission for an emergency closure to protect red abalone until a management plan could be adopted. Representing the Greater Los Angeles Council of Divers, Amy Anders testified that thousands of divers in southern California were "willing to give up their right to take abalone" in order to conserve the resource for future generations.[66]

On the other side, CAA president John Colgate argued that the state's environmental review should be completed before any moratorium was put into effect. CAA advocate Steve Rebuck said the situation with red abalone was different from other abalone species and urged consideration of alternative management strategies.[67] Other commercial divers insisted that their outplanting efforts had worked and again challenged the state's scientific studies as inadequate.[68]

When one commissioner asked the department to respond to public challenges about its science, CDFG senior biologist Dave Parker produced a stack of articles that Haaker, Karpov, Friedman, and others had published in peer-reviewed scientific journals. Haaker later told me that the state's abalone biologists, inspired by Tegner and the culture of science at the International Symposium on Abalone, had worked on their own time to analyze data and write up their research for publication.[69]

Though the commercial divers insisted that there were still enough red abalone at the remote islands to sustain continuing take, they had a boatload of science and history stacked against them. Resistance to regulation had been a harbinger and hallmark of many collapsing fisheries, from sardines and salmon to rockfish and cod, which famously crashed in the early 1990s, even as fishermen claimed there were still fish to catch.[70] Moreover, the commercial abalone divers carried on their backs the legacy of their predecessors, who had blocked meaningful reform in the 1970s.

The fish and game commissioners deliberated with discomfiture. No one wanted to cause hardship to commercial divers, but in the end, the prospect of doing nothing to protect California's abalone proved untenable. The commissioners voted unanimously to issue an emergency 120-day closure of the red abalone fishery south of San Francisco.

After the meeting, news reporters captured some immediate reflections and reactions. Commissioner Frank Boren explained: "It was too late to wait. ... With the white, black, green and pink, when we finally got the data on those populations, it was like looking at a heart attack victim that was already dead."[71]

The commercial divers were shocked. As longtime Channel Islands diver Kenny Schmidt put it, "I feel like I just got shot in the foot."[72] CAA secretary Michael Kitahara was infuriated that commercial divers had been vilified: "They've made it sound like we're the Exxon Valdez of fishing."[73]

That's when members of the Able Group started getting threats. Steve Benavides remembers getting an angry late-night call from a man who claimed to know his boat and warned that he'd better watch out. At the time, Benavides was heading out to Santa Barbara Island every three weeks to dive and take photos at a spectacular underwater cave. One afternoon, a five-hundred-pound black sea bass silently swam up from behind to hold right beside him. "I wasn't intimidated by the threats," Steve recalled, "but that giant fish scared the *beejeezus* out of me."[74]

* * *

That June, as the abalone bill moved to the state assembly, fish and game wardens on the North Coast made another big poaching bust at the Los Angeles airport—two repeat offenders ready to ship 1,700 abalones overseas, the largest single-day seizure in history.[75] The high-profile case kindled even greater support for the abalone bill.

The Greater Los Angeles Council of Diving Clubs hosted a booth at a big scuba expo aboard the *Queen Mary* in Long Beach and gathered dozens of personally signed letters from southern California sport divers that were mailed to assembly members, urging the abalone closure. Most added personal notes. One Long Beach diver's letter was typical: "I have been SCUBA diving for 12 years now . . . but I have seen less than a dozen abalones in all that time and I have never seen one that was legal size. I think it is time to do something."[76]

In the meantime, the CAA scrambled to develop an alternative plan. Because commercial divers wanted to keep Point Conception, San Miguel Island, and the Farallon Islands open for abalone take, they conceded to strict quotas based on the conservation strategy of maintaining minimum viable populations. They also urged the state to invest in experimental outplanting of abalone larvae rather than hatchery-raised juveniles, using the monies set aside in an account derived from landings fees they'd paid into since 1990.[77]

In July, the Assembly Committee on Water, Parks, and Wildlife considered suggestions of both sport and commercial divers as they made amendments to the senate bill. They boosted the abalone stamp fee, further increased fines for poaching, and finally required the department to develop a science-based abalone recovery and management plan, but ultimately kept the fishery closure for all abalone species south of San Francisco in place, voting eight to one in support.[78] In September, when Senator Thompson introduced a final conference bill to the full senate, it sailed through twenty-nine to eight, and Governor Wilson signed it into law the following month.[79]

California clearly needed to better steward its marine life, and abalone—once called the "California shell"—was a good place to start. It had taken an epic effort, with sport divers from around the state pulling together and tapping the expertise of courageous scientists to finally change course. With the fishery closed at the Farallon Islands and south of San Francisco and a clear directive for science-based management, a new chapter for West Coast abalone was set to begin. But was it too late?

PART 5
Imperiled
(1997–2019)

Now is different.

> —Dr. Jane Lubchenco, former director of National
> Marine Fisheries Service, 2007

It's very very tough to learn how to save an endangered species because so few are left.

> —Dr. Melissa Neuman, abalone recovery coordinator, National
> Marine Fisheries Service, *Daily Breeze*, August 28, 2016

It is estimated that one-third of all reef-building corals, a third of all freshwater mollusks, a third of sharks and rays, a quarter of all mammals, a fifth of all reptiles, and a sixth of all birds are headed toward oblivion.

> —Elizabeth Kolbert, *The Sixth Extinction: An Unnatural History*, 2014

The solution, of course, is to give yourself enough leeway—to leave enough sea between your ship and the shore so that even if the gusts blow you relentlessly toward the coast, you won't actually slam into it.

> —Bill McKibben, "A Little Leeway," *Orion*, July 2011

Endangered species and wild things in the remaining wild places need us to care for them not selfishly but selflessly, for their sake, the sake of everything and everyone who is not us, for the sake of beauty and all it implies. As we make our habitual appeals to practicality, the argument we cannot afford to ignore, the one that must frequently be on our lips, is this: We live in a sacred miracle. We should act accordingly.

> —Carl Safina, "The Real Case for Saving Species," e360.Yale.edu, 2019

CHAPTER 13
Racing Extinction

While the California Department of Fish and Game biologists were embroiled in meetings and paperwork related to the contentious red abalone fishery closure, National Park Service biologist Gary Davis took the lead on white abalone. Closing that fishery had been a critical step, but biologists knew it wouldn't be enough. Populations of pink, red, and green abalone along the Los Angeles and Orange County coasts from Palos Verdes Peninsula to Dana Point—closed to fishing for twenty years—had shown no signs of recovery. In 1997–1998, another massive El Niño brought its perilous storms and warm waters back to southern California. Seeing so few solitary white abs clinging to isolated rocks, miles from one another, continued to haunt Davis. If they were going to stave off extinction, a more proactive approach would be needed.

In 1998, Davis pulled together a group of marine biologists from different public agencies, universities, and private organizations in southern California, aiming to tap individuals with special expertise to quickly develop a recovery strategy and start raising funds. Recognizing the lack of state leadership, Davis hoped a private-public partnership might serve best to kick-start efforts to recover the white abalone. They called their group the Abalone Restoration Consortium.

At the outset, the expert group considered two approaches: either move far-flung, wild individuals closer together to foster reproduction in the field or start a captive breeding program. They recognized the risks of captive breeding—particularly disease and limited genetic diversity—but they decided it would be less risky to bring wild animals in than to move them, especially since aggregation was untested, and there had been no apparent natural reproduction since the late 1960s.[1]

Ultimately, the consortium developed a four-step plan. They'd survey historic habitat and locate survivors. They'd collect wild animals for broodstock. They'd breed and rear a new generation through mariculture. And finally, they'd aim to reestablish dense aggregations of adult white abalone in protected refuges so the animals could reproduce and begin to rebuild populations.[2] They hoped that getting the white abalone listed as a federally endangered species could help their efforts.

When Pete Haaker had recommended to his Department of Fish and Game supervisor that white abalone be listed as a California state "species of concern," he was told flat out that the state wouldn't do it. The underfunded agency didn't have the resources. The biologists knew they already had sufficient evidence to pursue listing the white abalone as a federally endangered species, so the group decided to provide all its research to the Seattle-based nonprofit Marine Conservation Biology Institute, which submitted a formal petition in the spring of 1999 to the National Marine Fisheries Service (NMFS), the agency charged with managing endangered marine species. The Center for Biological Diversity, an environmental group that specializes in endangered species advocacy, also submitted a petition to list the white abalone as endangered.[3] In response, NMFS designated the white abalone as a candidate species under the US Endangered Species Act (ESA) and contracted with Scripps Institution of Oceanography to conduct a formal review of the animal's biological status. Mia Tegner would be the lead biologist.

With the ESA listing under formal review, consortium members continued to advance their plan. Gary Davis invited NMFS biologist Dr. John Butler to join the group. Butler had been skeptical of the initial National Park Service (NPS) and California Department of Fish and Game (CDFG) population estimates. He'd questioned whether the agencies' submersible surveys had sampled enough habitat and looked in the right places. In fact, Gary had felt compelled to survey historic white abalone habitat *within* waters under Channel Islands National Park jurisdiction, since the NPS was shouldering most research expenses to that point. But the largest and last-known white abalone landings had come from the waters surrounding San Clemente Island, under US Navy control, and from more distant submerged deep reefs—Tanner and Cortes Banks. White abalone had also been landed in northern Baja, Mexico.

Butler pushed the group to conduct more-extensive surveys to get a more accurate population estimate and a better sense of how much suitable habitat was available, while locating and capturing animals for captive breeding. Since his agency was considering the white abalone ESA listing, Butler helped secure funds for another submarine cruise.[4]

In October 1999, a team of fourteen scientific divers from the NPS, CDFG, NMFS, UC Santa Barbara, US Geological Survey, and Scripps set out for a more ambitious set of submarine surveys that specifically included areas with the highest historic white abalone landings. They conducted seventy submersible dives and surveyed more than thirty-five acres at Santa Cruz, Anacapa, Santa Barbara, Santa Catalina, and San Clemente Islands and at the more remote Osborn, Farnsworth, Tanner, and Cortes Banks. This time, they found 157 live white abalones, but, confirming earlier worries, only eighteen (seven pairs and one group of four) sat close enough to reproduce. The overall white abalone population and density numbers were unnervingly low, and the effective breeding population was even smaller than previous estimates.

Now Butler was as alarmed as anyone. Based on the new sub surveys, the team estimated that fewer than three thousand white abalones remained, mostly on distant offshore banks. If nothing was done, these isolated, old animals would continue to die from natural causes, leaving no next generation to take their place.[5]

* * *

For the captive propagation effort, consortium biologists estimated they'd need to bring in at least two hundred wild white abalones to avoid inbreeding and preserve genetic diversity in offspring. But in all their surveys to date, they'd located only 172 animals, and most dwelled too deep for scuba divers to retrieve. By the end of 1999, they had brought in only one male.[6]

Redoubling efforts, CDFG divers managed to gather five more white abalones in March 2000 from near Catalina Island: this time, three females, a male, and one of unknown sex. But every step of the way, there were challenges. One night, an enormous school of squid surrounded their boat, clogging the seawater inflow to its live well. They had to turn off the pumps and ferry their precious abalone cargo ashore for emergency storage in an aquarium tank at USC's Wrigley Marine Science Center on Catalina. Then, they found they'd inadvertently injured some animals during collection.[7]

The next time, the crew set out with two vessels. On one, a team of NMFS scientists searched for abalone with a remotely operated underwater vehicle (ROV) that transmitted visual images from deep reefs. On the other, the CDFG dive team stood ready to make short, targeted bounce dives to retrieve the deep-dwelling animals located by the ROV crew. For ten full days, the ROV crew examined rocky habitat from sunup to sundown, yet they were able to collect only nine white abalones.[8]

Dr. Laura Rogers-Bennett, a marine biologist who had recently joined the Department of Fish and Game, later recalled the special protocol that divers developed to collect white abalone on those dives. When I talked to her at the Bodega Marine Lab (where she now holds concurrent research positions with both the renamed California Department of Fish and Wildlife and UC Davis), she took a white abalone shell down from her office bookshelf and handed it to me to demonstrate just how light and delicate it was—like porcelain compared with the sturdy stoneware of a red abalone shell. "First we'd dangle a piece of kelp to tempt the abalone to loosen its grip to the rock," she explained. "Then, we'd scoop it up with a soft plastic spatula to avoid injuring its fragile flesh." Ultimately, the department's divers were able to collect only eighteen animals.[9]

The consortium team didn't realize it at the time, but delays in endangered species permitting and the extreme scarcity of white abalone would preclude further collection for more than a decade.

✳ ✳ ✳

Meanwhile, Mia Tegner, with coauthor Dr. Alistair Hobday, completed the official White Abalone Status Review for the National Marine Fisheries Service, assembling known biological and population data. "Without protection and intervention," they soberly concluded, "the white abalone is likely to go extinct in California in ten years."[10]

Through her extensive field research with other abalone species, Tegner had been well poised to evaluate the white abalone's precarious condition, but she would not live to see the results of her work. In January 2001, while diving at a shipwreck off Mission Beach in San Diego, Tegner died in a scuba accident. She was only fifty-three—in the prime of her career as a marine scientist. Her husband, dive writer and underwater photographer Eric Hanauer, described the grievous accident with candor: "Mia was diving the way she had done thousands of times before. Although we may never know what really happened, I think she was so engrossed in her work that she didn't check her air and ran out." Hanauer added, "Sometimes we get so comfortable underwater that we forget we are riding a tiger."[11]

Ever since she'd first studied the possibility of outplanting abalone in the late 1970s, Tegner had continued to pay attention to the mollusks, even as she expanded her research to broader questions about the ecology of southern California's kelp forests. Urchins and abalone remained key protagonists in the compelling drama of ecological change she was tracking. When she helped organize the first International Symposium on Abalone in Baja in 1989, mentoring other West Coast abalone biologists to publish their research findings, her stature as America's top abalone biologist became established. Then, with her testimony to the California Fish and Game Commission urging the fishery closure to conserve abalone into the future, her reputation as a marine scientist who cared deeply was solidified. As a testament to her special role with abalone, an obituary article about Dr. Tegner in the *San Diego Union Tribune* credited her with "saving the abalone from extinction."[12] In 2005, the Department of Fish and Game would name a Marine Protected Area off Point Loma in her honor.[13]

13.1. Over several decades, Dr. Mia Tegner studied kelp forest ecology, including prospects for restoring abalone and the effects of El Niño oscillations on abalone growth and reproduction. She helped organize the first international abalone science conference in 1989 and advocated closing the commercial fishery in 1997 in order to conserve southern California's remaining red abalone. (Photo by Eric Hanauer)

Tegner's tragic death was an unspeakable loss to the community of scientific divers and researchers with whom she'd worked closely for more than two decades, as she'd incrementally pieced together the complex workings of southern California's kelp forests.[14] Through her colleagues, much of the research Tegner initiated would continue forward, illuminating the significant challenges increasingly besetting marine ecosystems.

Later that spring, based on Tegner and Hobday's status review, the National Marine Fisheries Service officially listed the white abalone as an endangered species. *Haliotis sorenseni* became the first marine invertebrate to earn such a dubious distinction.[15]

* * *

While the NMFS ROV team and the CDFG divers surveyed habitat and tried to locate and collect surviving animals, others started planning for the white abalone captive breeding program. Mariculturist Tom McCormick took a lead role. Founder and director of the Channel Islands Marine Research Institute (CIMRI), a private aquaculture research facility at Port Hueneme, McCormick brought practical abalone propagation expertise to the effort.

Plus, he offered the use of his local facility and another lab at Oxnard where the animals could be reared.[16]

Though mariculturists routinely propagated red and green abalone in tanks, white abalone had been raised experimentally only once before in captivity, thirty years prior.[17] The consortium tapped the expertise of UC Santa Barbara molecular biologist Dr. Dan Morse to crack the code of the white abalone fertility cycle and to develop a protocol for spawning and rearing the resulting larvae and juveniles. (Remember, Morse and his wife had discovered the molecular signals that induced red abalones' synchronous spawning and larval settlement back in the late 1970s.[18]) He found that white abalone gametes developed at a lower temperature than those of other abalone. Making adjustments to account for differences in environmental conditions and food, Morse came up with a new procedure for propagation.[19]

For security's sake, the wild white abalone retrieved in 2000 had been divided into two lots. Some were given to McCormick, who had tanks ready at CIMRI in Oxnard, and others were turned over to Morse and his collaborator Neal Hooker at UC Santa Barbara. Both groups of wild animals were carefully nurtured through the fall and winter in seawater tanks.[20]

By all accounts, spring was the one time of year when white abalones spawned. Sure enough, in late April 2001, McCormick identified a male and a female, each with swollen gonads. He packed them into a cooler and drove them north to Santa Barbara, where Hooker had two gravid females and a male ready to spawn. They set the animals into individual containers, dimmed the lights to mimic deeper water conditions, and added the requisite dose of hydrogen peroxide as an abalone "aphrodisiac" to induce spawning.

A couple of hours later, they watched as two of the females released golden sprays of gametes—about three million eggs each—through the multiple holes in their shells. Shortly thereafter, one of the males released a cloud of sperm. The biologists gathered the milky sperm and added it to the two batches of eggs. The mechanics of the meeting gametes was invisible to the naked eye, but with the aid of a microscope, they determined that more than 95 percent of the eggs were fertilized. The next day they had an estimated six million white abalone larvae swimming in their tanks, each smaller than the period printed at the end of this sentence. Overnight, they'd done a U-turn from the edge of extinction back to sturdier, more hopeful ground.[21]

McCormick and Hooker monitored the larvae carefully. After seven days, they figured that the tiny mollusks had reached the stage of larval development when they could be induced to metamorphose and settle. In the wild, abalone larvae, initially fueled by a small yolk, propel themselves up and down through the water column with minuscule cilia until they detect the chemosensory cue from coralline algae that indicates a promising place to settle. In the lab, the biologists added the inducer molecule gamma-aminobutyric acid (GABA) to a small sample of the larvae and observed as the animals began to settle out and metamorphose into tiny abalone.[22]

McCormick rushed most of the swimming larvae back to tanks in Oxnard, where he induced them to settle onto special plates coated with a thin film of diatoms that would provide food for the tiny animals' developing mouthparts. On call twenty-four hours a day, he kept close watch over the baby white abalones, mindful that the future of the species was riding on their survival. The earliest phases of an abalone's life are the most vulnerable, and natural mortality is extremely high. By July, only 7 percent of the initially reported millions had survived to reach a shell length of 2.3 mm. By September, roughly one hundred thousand baby abalones had survived the critical first three months of life and grown to the size of a ladybug—their tiny flat spiral shells already resembling the shells of adults.[23]

At that point, McCormick thought that the white abalone propagation program was off to a propitious start. The possibility of restoration looked more promising than ever, but unexpected challenges loomed.

＊ ＊ ＊

With the Endangered Species Act listing, the National Marine Fisheries Service hired a dedicated white abalone recovery coordinator, Dr. Melissa Neuman, who convened an official recovery team that included Haaker and McCormick from the original consortium group, plus a new generation of researchers from the National Park Service and the California Department of Fish and Game, as well as geneticist Dr. Ronald Burton from Scripps; veteran abalone mariculture expert Dr. David Leighton; and disease expert Dr. Carolyn Friedman, who had moved from CDFG's pathology lab to the University of Washington. Gary Davis, who had been long been a key player, shifted his focus toward establishing marine reserves—the refuges that would ultimately be needed to safeguard recovering abalone.

With Neuman, NMFS and its parent agency, the National Oceanic and Atmospheric Administration (NOAA), became critical partners, bringing additional capacity to the project through staff expertise and leadership, scientific divers, and ship- and ROV-research time, as well as funding in the form of grants and contracts to supporting organizations; but the ESA listing also brought some unexpected hurdles. Now all research and recovery activities required a permit. Also, as the nation's first ESA-listed marine invertebrate, white abalone challenged biological perceptions and pushed the boundaries of usual ESA-permitting protocols.

Typically, in considering permit applications, the ESA staff had to determine an acceptable incidental "take" based on how many animals could be expected to die from natural causes. They were most familiar with endangered animals like the California condor that produced a single chick, painstakingly nurtured from egg to adulthood. However, the broadcast spawning white abalone produced millions of eggs, only a small portion of which would survive to become juveniles, and only a tinier proportion of which could be nurtured to maturity in aquaculture tanks. Its natural mortality rate—99 percent—was unacceptably high by the standard ESA rubric. Ultimately, Haaker and Neuman educated federal regulators about abalone's reproductive biology and convinced them to make accommodations to approve the necessary permits.[24]

Neuman found tremendous enthusiasm for abalone in California, but when she went back to Washington, DC, to compete for federal funding against other endangered wildlife recovery efforts in the national arena, she realized that colleagues from other regions didn't get so excited about the big-footed marine mollusks with glimmering shells. Sea turtles, salmon, and whales had more charisma and tended to garner more attention and funding, owing partly to a historical hangover from the state legislature's 1913 decision to ban the export of abalone from California. In the Unites States, passion for abalone had remained mostly a California phenomenon. The lack of support owed also to environmentalists' "siding" with the sea otters back in the 1960s and 1970s. The primary abalone conservation advocates were not typical endangered-species champions with a broad base of citizen, media, and political support, but rather a small group of public-agency biologists trying to do right by the imperiled mollusk they felt entrusted to protect. Moreover, people who knew of abalone were still most inclined to think of them primarily as a wild food rather than as endangered animals in need of help.

Though hampered by limited federal funding, Neuman and the new recovery team pressed forward. Over the next five years, they would develop an official recovery plan that built on the consortium's initial approach but also addressed new threats—most important, the susceptibility of captive white abalone to disease.

* * *

Soon after McCormick settled the white abalone larvae in his tanks in 2001, he began research, raising different sets of animals under different conditions to determine which water temperature and diet was best for growth. The one previous study of raising white abalone via aquaculture had indicated that warmer temperatures, around 64 degrees F, sped the animals' growth. However, after five months, McCormick discovered that juvenile abalone growing at that warmer temperature were turning up on the bottom of his tanks, upside down and unable to right themselves—a telltale sign of withering syndrome. The disease bacteria had apparently infiltrated when a UV-filtration unit malfunctioned and failed to properly sterilize the seawater flowing into the aquaculture tanks.[25]

McCormick immediately contacted shellfish disease expert Carolyn Friedman. At the time, she was working with the Abalone Farm in Cayucos to develop an antibiotic feed that could safely ward off withering syndrome in mariculture-raised, cocktail-sized shellfish destined for market. The farm quickly ferried some of its antibiotic kelp feed down to Oxnard, where Mc-Cormick was able to treat the sick white abs and rescue many from the most damaging effects of the disease.

But the larger picture looked grim. McCormick first observed signs of withering syndrome in late September 2001. By early November, half the animals had died. By the following February, only 8 percent remained. Since the numbers had been so high to begin with, there were still thousands of juvenile white abalone clinging to life, but the downward trend was worrisome. By lowering the water temperature, McCormick kept symptoms in check, but the small mollusks still harbored the deadly disease bacterium.[26]

The consortium's greatest fears about disease were coming to pass. The specter of withering syndrome clouded the prospects for breeding white abalone in captivity. Even more troubling, since the pathogen was now considered endemic in all the waters of southern California, where could the captive-raised white abalone ever be outplanted? Some biologists remained hopeful that the animals could survive if returned to the cold refuge of their deepwater habitats. But others were concerned that if researchers inadvertently outplanted a diseased individual, the bacteria could spread to remaining wild animals, squelching the remote yet wishful possibility that wild white abalone, if left alone, might someday stage their own comeback.

The NMFS-led recovery team nixed the consortium's original plan to outplant captive-bred animals until disease problems could be brought under full control. Ultimately, Friedman and her successor at the CDFG, Dr. Jim Moore, developed antibiotic bath treatments that completely cleared the withering syndrome infection in captive animals.[27] McCormick successfully raised another small batch of white abalone in 2003, but given persistent difficulties with water quality, he decided not to renew his ESA permit.

In 2008, for security's sake, the recovery team divvied up the remaining captive-bred animals—only fifty individuals—to be held at different aquarium facilities, and decided to move the white abalone captive propagation effort north to the University of California's Bodega Marine Lab, where the seawater remained clean, cold, and free of withering syndrome.

UNEXPECTED PERILS: BLACK ABALONE

By the time the NMFS had designated the white abalone as a candidate for the endangered species list in 1997, it had become clear that the intertidal black abalone warranted consideration for listing, too. Withering syndrome had appeared in new spots on the mainland coast, beyond the artificially warmed waters of Diablo Cove. In 1994 it reached south to the coast off Vandenberg Air Force Base near Point Conception. By 1997, it had spread north to the San Luis Obispo County line. And by 1998, it had reached Cayucos Point, abetted by warm water from the severe El Niño that winter (1997–1998). In addition, the disease ravaged commercial red abalone farms on the Central Coast, killing most of their animals.[28]

Meanwhile, disease specialist Carolyn Friedman had proceeded with research on the virulent pathogen and homed in on a key microbial suspect—a *Rickettsia*-like organism related to Rocky Mountain spotted fever that seemed to be transmitted through the animals' feces.[29] The offending microbe, though relatively common and innocuous, in the particular case of abalone infected the mollusk's digestive tract and thoroughly disrupted nutrient uptake, compelling the animal to metabolize its own foot muscle, ultimately with lethal result.[30]

Given its devastating effects on black abalone, the withering syndrome bacterium behaved like a nonnative invader, but no similarly injurious abalone pathogen was known to exist anywhere else in the world. According to Dr. Kevin Lafferty, a US Geological Survey researcher who specializes in the ecology of parasites and pathogens, it is not uncommon for pathogenic organisms to coexist with their hosts in one ecological context, where animals are well adapted, but then to wreak havoc in a new place under new environmental conditions. Lafferty thinks the disease could possibly have come in with ballast water, with an influx of warm El Niño water, or with imported shellfish. The accidental introduction of shell-boring sabellid worms from South Africa through commercial abalone farms in the mid-1990s revealed how readily parasites and pathogens could globe-trot and spread.[31] Moreover, with the poorly tracked outplanting of hundreds of thousands of hatchery-raised juvenile abalone by all manner of well-meaning people—from CDFG biologists to commercial divers, sport diving clubs, college classes, civic clubs, and municipalities—there were numerous scenarios by which an exotic pathogen could have been inadvertently introduced, and then amplified and spread in the seawater warmed by California's mammoth 1982–1983 El Niño.[32]

Ongoing research by Friedman and her colleagues soon confirmed that all species of abalone in California were susceptible to withering syndrome to one degree or another, especially during warm water periods associated with El Niños, making it clear that plans to restore any abalone in southern California would now need to overcome this invisible but formidable bacterial obstacle.[33]

With the disease continuing to march north up the coast, NMFS designated the black abalone as a prospective "candidate" for Endangered Species Act listing in 1999 and began to orchestrate research necessary to formally evaluate the animals' status.[34] With white abalone, the ad hoc group of consortium biologists had surmounted the difficulties of "listing" a broadcast spawning mollusk for the first time and began to develop strategies to recover the species. Circumstances and challenges for the intertidal black abalone, it turned out, would be very different from those of their deep-dwelling subtidal cousins.

* * *

Commercial divers had raised concerns back in 1992 that closing the black abalone fishery would hinder research, but the withering syndrome epidemic had, in fact, attracted the interest of a legion of new researchers who recognized the disease's significant ecological repercussions. Despite the CDFG's severely limited resources for studying black abalone, several other government and academic researchers stepped in to fill the gap. The department's past abalone research had focused on parameters specifically needed to manage the fishery, such as growth and recruitment rates; however, much of the new research expanded to also include population structure and ecology. By the late 1990s, results from more diverse and robust abalone research began to unfold and afford new insights into the mollusks' plight.

Long before withering syndrome hit, marine researchers had begun to inventory and monitor intertidal sites on California's mainland coast in response to growing concern about potential oil spills, with increased tanker traffic and the ever-looming possibility of offshore drilling platform blowouts, as had occurred calamitously in the Santa Barbara Channel in 1969. With the 1989 *Exxon Valdez* debacle in Alaska, the federal government realized that to accurately assess monetary "injury" for losses following a disastrous oil spill, it needed to better understand what public values—including fish and wildlife, plus nursery, feeding, and rearing habitat—were present beforehand. With funding for this task from the Department of the Interior's Mineral Management Service, UC Santa Cruz researcher Dr. Pete Raimondi established seven intertidal monitoring sites—from southeast of Point Conception to north of San Simeon—and was well positioned to observe black abalones as the withering syndrome epidemic spread both south and north from Diablo Cove. He'd eventually add more monitoring sites north through Big Sur to Point Lobos.[35]

As the disease advanced, Raimondi and his colleagues witnessed the dramatic die-off at each of their Central Coast study sites. But that wasn't the end of the story. After the black abalone died, Raimondi observed that encrusting tubeworms quickly took over, filling up all the empty crevices and cryptic spaces. These were the protected nursery habitats that juvenile abalone needed to settle and grow. Even more important, the tubeworms smothered pink crustose coralline algae that provide the key chemosensory cue for abalone's larval settlement. When I talked with Raimondi about his research, he explained that he'd studied molecular biology at UC Santa Barbara with Dan and Aileen Morse back in the 1970s—when they'd made the landmark discovery that coralline algae emitted a GABA-like molecule that cued planktonic larval abalone to settle down and develop into juveniles. Now, in his own field research, Raimondi could see the profound ecological importance of that insight.

Without adult black abalones actively grazing the surface of the coralline algae and keeping it clear, it was becoming more difficult for a next generation to gain its footing. Raimondi had elucidated a second kind of Allee effect. Not only did the withering syndrome epidemic drastically reduce the number of adults below the threshold needed to reproduce, but it also left too few adults to maintain the pink algal surfaces that the larvae needed for both settlement and food, presenting yet another obstacle to the animals' natural recovery. Adding insult to injury, the encrusting tubeworms also preyed on abalone larvae.[36] "If all the adult black abalone die," Raimondi told me, "then there is a negative feedback loop; it's basically a death spiral for the local population."[37]

∗ ∗ ∗

Meanwhile, for its official black abalone "status review," NMFS needed to evaluate the population structure of the species. Were black abalone from the animal's entire distribution range genetically similar, or did localized aggregations have distinct genetic profiles? The answer to this question was crucial for determining not only the status of the species across its entire range but also whether animals from still-healthy but distant northern populations could someday be transplanted to help repopulate southern areas where black abs had already been locally extirpated by withering syndrome.

Because there was no good method to directly track the dispersal of microscopic abalone larvae, biologists looked to genetic markers to infer dispersal patterns and to determine whether or not populations were closely related. Scripps geneticist Ronald Burton led the research. With the help of CDFG biologists, he obtained small tissue samples from four hundred black abalones at seven different locations on the Central Coast and then analyzed gene sequences. (Though abalone are highly vulnerable to cuts on their feet, it turns out that the tips of their epipodial tentacles can be clipped without risk.)

Burton found that black abalone from the same sites shared similar genetic traits, but animals from different sites had distinctly differentiated genes. This pattern indicated that there was little gene flow between the various isolated populations and that most black abalone grew up from larvae produced by local parents, research that was affirmed by additional genetic studies. Over long periods of time, there might be occasions when larvae could travel farther, resulting in a greater mix of genes, but that was the exception rather than the rule.[38]

Without adults to provide a nearby source for new larvae, there was little possibility that black abalone could naturally reseed themselves from more distant areas. Considered together with Raimondi's findings about how rapidly tubeworms smothered intertidal nursery habitat, prospects for black abalone recovery looked grim. Once a local population disappeared, there would be no easy way to bring it back.

* * *

With white abalone recovery efforts blazing a path forward, researchers at UC Santa Barbara wanted to start a captive breeding program for blacks, too, but the intertidal mollusks were considered notoriously difficult to spawn. In 2007, Dr. Hunter Lenihan re-created intertidal conditions in special tanks aiming to prime captive black abalone for reproduction, but the males showed little sign of gonad development. When the late-summer spawning period arrived, he nevertheless experimented with a variety of stimuli, including elevated temperature and hydrogen peroxide—cues that reliably set off spawning in other *Haliotis* species—but with no success. Ultimately, he had to abandon the project.[39]

Meanwhile, with studies of distribution, disease, propagation, and genetics ongoing, the Center for Biological Diversity, a group that watchdogs ESA implementation, grew impatient that NMFS had not yet completed its status report for black abalone and, in 2006, submitted its own petition for listing. The external petition triggered a legal requirement for NMFS to respond within a ninety-day time frame, forcing the review onto a fast track.[40] The agency's final status report concluded that, if nothing was done, "black abalone were likely to become effectively extinct in 30 years." In 2009, recognizing disease and elevated seawater temperature as grave threats to the mollusk's survival, NMFS officially listed the intertidal black abalone as a federally endangered species.[41]

Now, two of California's seven abalone species were endangered and under federal protection.

CONTESTED PLANS: RED ABALONE

The 1997 legislation that closed California's red abalone fishery south of San Francisco also directed the California Department of Fish and Game to develop an abalone recovery and management plan for all the state's *Haliotis* species. At the time, few small-scale fisheries had management plans, but the need for reform was becoming widely recognized. Just the year before, to address mounting concerns about overfishing nationwide, the US Congress had enacted significant amendments to federal fisheries law (the Sustainable Fisheries Act, an amendment to the landmark Magnuson-Stevens Fishery and Conservation Management Act of 1976) and directed regional fishery management councils to take a more precautionary approach.[42] Then, in the following years, with mounting support for ocean conservation, California's legislature enacted the Marine Life Management Act (MLMA), requiring the state to develop plans for sustainable management of all its fisheries, and the Marine Life Protection Act (MLPA), requiring the state to develop plans for a network of Marine Protected Areas up and down the coast.[43] The state's fisheries biologists and managers were excited by the visionary new directives and abuzz with figuring out how to do a better job. Since the state's abalone plan mandate had come first, it would be on the cutting edge.

The rationale behind new fishery management plans was to set clear, science-based criteria to guide and trigger specific actions, with the aim of eliminating the political interference that had chronically hampered the ability of the CDFG to respond quickly to warning signs.[44] Abalone had been a case in point, with commercial divers opposing restrictions to their fishery every step of the way. Moreover, the MLMA directed the department to engage stakeholders ahead of time, to give everyone an opportunity to provide input on how the plans would best work. The hope was that publicly vetted science-based plans—with buy-in from fishermen up front—would ultimately lead to better outcomes for fished species, especially when the chips were down.

However, there was an important difference with the Abalone Recovery and Management Plan (ARMP). As state biologist Pete Haaker later explained, "Not many other fisheries were in such bad shape that they had to be *recovered* first."[45] Since the commercial abalone fishery was closed, the department re-garded the ARMP foremost as a blueprint to recover depleted populations in southern California while also managing the still-robust sportfishery for red abs in northern California.

Developing the plan afforded the state biologists an unprecedented op-portunity to pull together data from landings, surveys, and decades of abalone research from around the world to inform new conservation and restoration strategies. Realizing that reliance on landings data had contributed to the serial depletion of stocks, they wanted to base the new management plan on their best understanding of abalone biology and fishery-independent data.[46]

Ultimately, they decided on a strategy that emphasized tracking density.[47] Following up on concerns about the Allee effect, New Zealand and Australian researchers had determined that, when reproductive male and female abalones sat 1.6 meters apart, their gametes had only a 50 percent chance of fertilization, highlighting the need to tighten up "nearest neighbor distances."[48] As such, the ARMP specified a two-thousand-animal-per-hectare criterion (equivalent to two animals per 10 square meters) as the minimum density for recovery—or the minimum viable population. But only when abalone reached a density of 6,600 animals per hectare (six per 10 square meters) would the department consider reopening a commercial fishery. Because there was not yet sufficient data to develop recommendations on a species-by-species or location-by-location basis, the general density criteria would apply to abalone statewide until additional research allowed for greater refinement.[49]

The CDFG biologists aimed to follow the "precautionary principle" that had finally gained broad credence in natural resource and fisheries manage-ment. In the face of uncertainty and in the wake of staggering population declines, they sought finally to err on the side of conserving abalone.[50]

For northern California, where red abalone remained plentiful, the chief goal of the ARMP was to continue managing the popular sportfishery. State bi-ologists identified a total allowable catch (TAC), based on past take, and then divvied it up into bag limits. The CDFG would keep tabs on the abalone popu-lation by monitoring density and recruitment. CDFG environmental scientist Laura Rogers-Bennett started a long-term kelp forest monitoring program for the North Coast to track abalone and also competitors and predators so the department could gain a better understanding of ecosystem dynamics.[51] The agency's biologists also developed a new method using "abalone recruitment modules"—basically small piles of cinder blocks in prime habitat that could be pulled apart to determine and track how many new animals joined the population each year. In addition, department biologists had improved ways to monitor changes in fishing pressure by tracking "catch per unit of effort"

(CPUE). If telephone creel surveys indicated that sport divers were spending more time to take their bag limit, it would trigger dive surveys in high-use areas to field-check densities. If abalone density fell below the critical threshold, it could then trigger closures to conserve broodstock in localized areas.[52]

For central and southern California, the ARMP's goals were to prevent further depletion, reverse declines, and then rebuild self-sustaining abalone populations that might someday support a renewed fishery. State biologists were frank that this final goal could take decades. The plan identified possible methods to repopulate empty habitats, such as translocating and aggregating adults and outplanting larvae, though the actual feasibility and cost-effectiveness of these approaches remained uncertain. A 1997–1998 El Niño that surpassed the severity of the 1982–1983 warming event had again knocked back abalone recruitment.[53] Moreover, the prevalence of withering syndrome and the potential expansion of sea otters into southern California waters still presented formidable obstacles to bringing back the surplus of animals needed to support a fishery.[54]

Department biologists worked with a scientific advisory board to put their draft plan through a formal peer review, then, in 2003, hosted workshops and more than a dozen meetings to garner input from the public, including the same sport and commercial divers who had long been involved with abalone issues.[55] Although a new generation of state biologists intended to make the ARMP a fresh start—a science-based, forward-looking blueprint—the public process invariably carried with it the same baggage of conflict, animosity, and distrust that had permeated the "abalone war" of the late 1990s.

Not surprisingly, commercial divers became the plan's most vocal critics. Undaunted by risks of further depletion or extinction, they argued that the ARMP didn't reopen areas to commercial fishing fast enough. By 2003, when the draft plan was first put out for public review, the red abalone fishery had been closed for five years—a time-out that had allowed populations at distant San Miguel Island to somewhat rebound. The commercial divers contended that one submerged reef, a place they called "the Miracle Mile," now had ample shellfish to supply a renewed commercial harvest.[56]

The Marine Life Management Act had encouraged broader participation of commercial fishermen in management of the state's fisheries, and so the California Abalone Association (CAA) hired Australian researcher Dr. Jeremy Prince to help them refine a new approach to micromanage the small area as a sustainable, artisanal fishery. Prince had worked with commercial abalone divers in Western Australia to develop different methods of abalone management based on reef-scale surveys conducted by divers with a focus on maintaining local spawning stock and exclusive, private-rights-based access to specific areas that created financial incentives to avoid overfishing.[57] The CAA developed an independent proposal for an experimental commercial fishery at San Miguel Island based on a quota system. With a high level of distrust of the CDFG biologists, the commercial divers proposed to collect their own data and hire their own biologist to manage the fishery with oversight by the Fish and Game Commission. To counter the CDFG's contention that it didn't have the resources to monitor such a small commercial harvest and to avert poaching, the commercial divers proposed to conduct their own monitoring and enforcement.[58]

In November 2005, commercial diver Chris Voss presented the CAA experimental fishery proposal to the Fish and Game Commission. (The MLMA had also transferred oversight of commercial fisheries from the California legislature to the commission.) It was one month before the Department of Fish and Game was scheduled to finally present its publicly vetted Abalone Recovery and Management Plan. By then, the commissioners who had presided over the harrowing hearings that closed the black, white, pink, green, and red abalone fisheries in the 1990s had termed out. The newly appointed set had not heard all the recreational divers' laments or Dr. Mia Tegner's compelling testimony about the need to conserve San Miguel's stronghold of abalone broodstock, so they were receptive to the commercial divers' fisherman-driven approach. Although recreational dive clubs strongly opposed opening San Miguel Island, they were the latecomers, this time. Cen-Cal president Steve Campi voiced surprise and concern that the commission would even consider the proposal seriously. However, the new commissioners—interested in engaging with the commercial divers in a "private-public partnership"—requested that the department add the option of opening an experimental fishery at San Miguel Island to the ARMP.[59]

The department added the option in deference to the commission's directive, but at the next commission meeting, when staff presented the science-based Abalone Restoration and Management Plan, they strongly recommended against opening any new fishery before the plan's recovery criteria were met. Department biologists wanted to follow precautionary management—as required by the state's new marine fisheries laws, especially given rising concerns about disease and climate change. Above all, they wanted to prevent remaining abalone populations from falling further below critical thresholds.

However, in response, Commissioner Jim Kellogg openly revealed his dim view of science, saying he'd accept the word of fishermen who make their livelihood from the sea "over all of the science in the world."[60] The other new commissioners viewed the commercial divers' private-rights-based proposal as trying something new and innovative that might become a model for other struggling fisheries—not in the biological context of abalone's vulnerability or in the larger historical context of political interference that had put the animals at such high risk in the first place.[61]

The department's Law Enforcement Division chief warned that reopening a commercial market for abalone would once again increase the likelihood and risks of poaching. Nevertheless, at the urging of the commercial divers, the Fish and Game Commission adopted the final ARMP with the possibility of opening an experimental commercial abalone fishery at San Miguel Island.[62]

Recreational divers who had worked to conserve abalone as a public resource in the late 1990s were outraged. Steve Benavides considered the addition of a fishery option "an unbelievable tragedy."[63] "To allow commercial or recreational harvest of a devastated species in one of the last holdouts of significant populations," he wrote, "is madness and greed personified."[64] Meanwhile, commercial abalone divers regarded the experimental fishery directive as a major victory—a way to possibly continue their livelihoods with a fresh opportunity to do things right. "None of us want to operate a fishery the way we used to," said former abalone diver Jim Marshall, who, like many, had switched to primarily harvesting red urchins. "We want a sustainable fishery that can operate over a long time."[65]

Many had hoped the landmark ARMP would finally give the department a way to focus on abalone recovery. The plan had outlined an extensive docket of research, including studies to determine the feasibility of various restoration methods, especially for pink and green abalone in southern California. However, as it turned out, the commission directed department staff to first study and collaboratively develop options to reopen the commercial fishery at San Miguel, leaving the other projects to founder.

* * *

The biggest bone of contention regarding the reopening of San Miguel Island was how many abalone were actually out there. Commercial divers, who regularly harvested urchins in the vicinity, claimed there were a lot, but agency biologists conducting transect surveys found too few. To head off potential disputes, Ian Taniguchi, who had succeeded Pete Haaker as abalone biologist at the CDFG, orchestrated a collaborative three-year dive survey, bringing together dozens of commercial ab divers, sport divers, and scientific divers from Channel Islands National Park and UC Santa Barbara to gather data to establish parameters for a targeted fishery. Meanwhile, the department convened a new Abalone Advisory Group with a more diverse set of stakeholders, including commercial and recreational divers, academic scientists, and, for the first time, marine conservation organizations. The group met regularly to consider how a new San Miguel fishery might be managed.[66]

Although the goal of the collaborative surveys was to create one set of shared data that everyone could agree on, the group ultimately disagreed about what the data meant. Relying on the density criteria specified in the ARMP, the biologists determined that there weren't enough abalone to sustain a fishery, while the former commercial ab divers argued that density wasn't a good measure to account for the patchy and linear configuration of San Miguel's abalone, which tended to congregate along reef edges. The Abalone Advisory Group developed four possible scenarios but couldn't come to any clear consensus. The CAA advocated a cooperative commercial fishery with commercial divers getting 90 percent allocation (they figured sport divers already had access to the entire North Coast), but the recreational divers, conservation organizations, and agency scientists favored no-take or restoration options, such as transplanting some of San Miguel's red abalone to empty habitat elsewhere, until recovery criteria were met.[67] In 2010, with the stakeholder advisory group unable to make a consensus recommendation, the Fish and Game Commission, with some new members, punted, directing the department to proceed with more modeling, cost estimates, and risk analyses.[68]

Following another period of poor recruitment and evidence of more withering syndrome, those analyses provided no greater reason for agreement. Moreover, a bust of commercial urchin divers poaching abalone from San Miguel Island reduced trust and further soured potential for consensus.[69] Ultimately, in 2013, after seven years of consideration—and more than $1 million spent—the Fish and Game Commission decided against opening the San Miguel experimental fishery.[70] Commercial abalone divers who'd invested money and effort into the possibility of opening the targeted fishery were disgruntled, but those who favored abalone conservation were relieved that San Miguel Island would remain a stronghold for red abalone broodstock in southern California.

Though the fish and game commissioners declined to put San Miguel's red abalone at any heightened risk, they wanted to find a way to address concerns raised by the commercial divers about how the department was evaluating remaining abalone populations for their future fishing potential. As a way forward, they directed the department to engage in an independent scientific review to update its ARMP approach—already nearly a decade old—and to start developing a new fishery management plan (FMP) for the state's recreational fishery in northern California.[71]

With so much attention focused on the project of opening a small commercial abalone fishery at San Miguel Island, it was easy to lose sight of a larger and far more significant issue: the ocean at coastal California's doorstep was beginning to turn into a cauldron of bouillabaisse.

RISING TEMPERATURES

Before she died, marine ecologist Mia Tegner had urged her Department of Fish and Game colleagues to consider seawater temperature as a parameter in the emerging Abalone Recovery and Management Plan to better account for the crucial role of environmental variation in abalone reproduction and recruitment.[72]

As part of her research to tweeze apart the factors that influenced southern California's kelp forests, Tegner had homed in on the pivotal role of seawater temperature. Through the course of her life's work, she had conducted research related to three major El Niños, observing firsthand how elevated temperature conditions depressed abalone recruitment, knocked out food sources, and amplified withering syndrome. Tegner's research had convinced her that California's broader abalone collapse had resulted from a series of stacked-up environmental stresses topped off by intense, unsustainable fishing pressures.[73]

There had been no way for state abalone fishery managers to nimbly lower fishing pressure during periods of poor recruitment and reduced kelp availability. In fact, during the high-stress period after the 1957–1958 El Niño, the state had actually loosened restrictions to keep up the catch—in retrospect, exactly the wrong response.[74] It would be far better for abalone if the state had authority to temporarily close a fishery when environmental conditions deteriorated.[75] The science relating temperature variation to fisheries and ecological productivity was not sufficiently developed to inform the ARMP in 2003, but it was advancing rapidly—and alarmingly.

By the late 1990s and early 2000s, the broader scientific community had amassed distressing evidence about the ominous perils of global heating, and oceanographers had already projected additional increases in sea surface temperatures for southern California. Mia Tegner realized that the historic El Niño pulses of intruding warm water were proxies—essentially postcards from the future—that could help biologists understand what larger warming trends might mean for abalone and other kelp forest denizens.

After her untimely death, several of Tegner's colleagues followed her line of inquiry with more focused research on how temperature variation could affect abalone. In one lab experiment, they subjected red and green abalones to different temperature regimes—similar to natural conditions for cool, normal, and warm phases of the California Current in the Southern California Bight—and found that elevated temperature affected the animals at every phase of life: larval settlement was impaired, growth was limited by lack of food or some yet-to-be-determined metabolic change, and survival of juveniles was hampered. In particular, researchers determined that red abalone suffered far more under warm water conditions than green abalone, which preferred warmer water conditions to begin with. Though elevated temperature exacerbated withering syndrome for most abalone species, researchers found that green abs infected with the disease and exposed to warm water fared much better than reds.[76]

Another lab experiment indicated that sperm production in male red abalones was highly sensitive to water temperature—akin to the "hot tub effect." Male gonads did not properly develop to produce viable sperm under warm conditions. A related experiment on females found that reducing food availability—a factor also influenced by temperature since warm water conditions shrink kelp supply—precluded the development of mature eggs.[77]

All in all, the emerging body of research showed that warming ocean temperatures not only amplified the virulence of withering syndrome but also had subtle effects on animals' development, growth, and survival that could wreak havoc on whole populations—especially important considering predictions for more-frequent and longer-lasting warm water conditions in the future.[78] In multiple ways, warm water conditions could knock back red abalones' ability to reproduce—even before their gametes ever had a chance to meet.

And there was more. Beyond the water-warming effects of climate change, oceanographers were also tracking alarming shifts in ocean chemistry. With excess carbon dioxide accumulating in the atmosphere, much of it was dissolving into the ocean, and rapidly tilting the pH of the seawater toward acid—the phenomenon known as ocean acidification. The increased acidity dissolves and fundamentally lowers the amount of calcium carbonate in solution in the ocean—like the inverse of the familiar Tums antacid tablet.[79]

Since calcium carbonate is a key building block of the shells and bones of the vast majority of marine life, researchers became increasingly alarmed that the downward shift in pH could have severe and consequential impacts on the basic biology of countless organisms and on the ecology of the ocean. In lab experiments and in high-latitude ocean waters, they had already observed that calcifying organisms—including shelled mollusks and reef-building corals—were showing a marked decline in their ability to make shells and skeletal structures in water with elevated acid.[80] In fact, the shells of planktonic pteropods (free-swimming snails with "winged feet" commonly known as sea butterflies) in both the North Pacific and Southern Ocean, where acidic conditions had already become pronounced, were already showing signs of pitting and dissolution. It was becoming impossible for the minute animals to keep up with building their delicate shells in the ambience of acidified water.[81]

For broadcast spawning marine animals that relied on ocean water as medium for both fertilization of their gametes and development of their larvae, rising acidity posed even more fundamental challenges. Upwelling of water with insufficient carbonate was already affecting oyster farms in the Pacific Northwest. Starting in 2006, aquaculture farms had repeated and dramatic failures with larval recruitment and viability of juvenile oysters, eventually attributed squarely to the corrosive conditions of increasingly acidic seawater.[82]

Researchers in Australia were the first to conduct lab experiments focused on how ocean acidification could affect abalone in particular. In many cases, larval animals under acidic conditions developed abnormally and did not build protective shells.[83] Follow-up research by Canadian researchers on pinto abalone (*Haliotis kamtschatkana*, listed as a "species at risk" in Canada in 2003 and as "endangered" by the State of Washington in 2019) found similarly disturbing results.[84] Japanese scientists determined that lower seawater pH associated with elevated CO_2 also sabotaged the fertilization of abalone gametes and increased mortality at the earliest stages of larval development.[85]

Meanwhile, adding to the alarming evidence, researchers determined that acidifying waters also had grave effects on crustose coralline algae—the pink, biotic substrate upon which all species of abalone depend both for chemosensory cues that trigger larval settlement and for nursery habitat at the very earliest stages of benthic life.[86] Like mollusks, it, too, builds its hard structure from calcium carbonate in seawater. Even as hard-won protections from fishing pressures provided a long-needed reprieve for southern California's abalone, it was becoming clear that ocean warming and acidification had potential to pile on and devastate the mollusks in many other ways.

By many predictions, the twenty-first century Pacific Ocean will become warmer, stormier, more acidic, and far less habitable to California's abalone. The already endangered white and black abalones, plus all the species of concern—pinks, greens, pintos—and increasingly beleaguered reds, will face greater challenges. Only by protecting as many animals as possible—to secure a Noah's Ark of genetic diversity—can we have hope that some abalone will harbor in the deep memory or mysterious mutability of their genes the capacity to be resilient and to endure the predicted and unknown trials ahead.

That hope lies at the center of the last chapter of the extraordinary history of California's most iconic and imperiled shellfish.

CHAPTER 14
Against All Odds

Dr. Kristin Aquilino, director of the University of California's Bodega Marine Lab's white abalone captive breeding program, remembers the day—the moment, in fact—when she happened to have a pipette in her hand and was able to reach down at the critical moment to suck up a dilute dose of *Haliotis* sperm.

It was June 2012. It had been nine years since Tom McCormick last spawned white abalones in captivity. The propagation program had languished owing to severe funding limitations and permitting delays, plus aging and deaths of the wild-caught broodstock animals, but finally Bodega Marine Lab (BML) had upgraded its seawater filtration system and obtained the needed Endangered Species Act permit, and a new generation of researchers was ready to move ahead. That was Aquilino's first year on the job.

Starting in April, the BML captive breeding team traveled to southern California three times, aiming to marry the gametes from white abalone housed at four different facilities—paired aquaria in Long Beach and San Pedro (Aquarium of the Pacific and Cabrillo) and in Santa Barbara (UCSB and the Santa Barbara Museum of Natural History's Sea Center). But the results were disheartening. On the first attempt, three females but no males spawned. The second time, females and males spawned, but the eggs were few and of poor quality. In two other attempts, females again cooperated but the males did not.

It was almost the end of spawning season, but the BML team traveled south again in mid-June to try spawning the fifteen white abalones held at UC Santa Barbara. At 5 a.m. they'd started the protocol—examining animals, assessing their gonads, and putting them into separate buckets with the "aphrodisiac" hydrogen peroxide solution. Then, they waited. Several hours later, two females expelled eggs. They waited some more, expectantly watching the lethargic males. That's when Aquilino noticed the weak spray of sperm, immediately retrieved it and mixed it with the waiting eggs. About three hundred thousand were fertilized. They packed the fresh embryos into tubs, put them on ice in the van, and headed north over Gaviota Pass, hurrying back up Highway 101 to the safe haven of Bodega Marine Lab's tanks before the microscopic animals developed into their fragile larval stages and prepared to metamorphose and settle.

It was a fresh breath of hope for the white abalone recovery effort—the first successful spawning in more than a decade. Aquilino recalls she was a "helicopter parent," watching over those microscopic mollusks as they passed through the difficult bottlenecks of early development and began to grow on diatom-coated plates. By the year's end, the team had added twelve new white abalones to the small captive population.[1]

Over the next few seasons, the BML team managed to boost production by an order of magnitude each year—raising 120 animals from spawning efforts in 2013 to more than 2,000 in 2014. By 2015, they had propagated enough juveniles to divvy them up to partnering universities and aquaria, with the aims of distributing risks and ramping up production. To start establishing even just a few functional white abalone populations, ecologists determined the program would need to reliably produce at least five thousand animals for outplanting each year, over the course of many years.[2]

"When people started the white abalone restoration project," Aquilino reflected, "I think there was a sense that it wouldn't be that difficult. After all, aquaculture farms routinely raised red abalones." In fact, it turned out to be very difficult, because the endangered white abalone recovery program faced serious constraints that farms don't. "When we spawn the animals, they give us all the gametes they have," Aquilino explained, "but those gametes aren't as plentiful or robust as they should be." Back in 2001, when McCormick and Hooker first spawned white abalone, the animals were ripe and ready from the ocean, and females produced three million eggs each; but at BML, with captive abalones sitting in filtered-seawater tanks for more than a decade, the females produced only about three hundred thousand. Likewise, torpid males were often recalcitrant or produced just a wisp of sperm.

Aquilino and other BML researchers suspected that there were natural cues that prompted abalone in the wild to put energy into preparing their gonads to fully ripen—cues that were lacking in captivity—so in 2013 they created a special brooding area where they could adjust seawater temperature and mimic southern California light, aiming to approximate ambient conditions of white abalone's natural habitat. To better prepare the abalones for spawning, the BML team also fed the mollusks a special high-protein, red-algae diet. BML researchers also stayed laser focused on maintaining the health of all the animals, continually testing for any sign of disease and waxing the mollusks' shells to protect against boring organisms.[3]

However, another constraint hampered the captive breeding program. The small number of reproductive adults not only made it difficult to cajole enough males and females to produce gametes at the same time but also limited genetic diversity, dimming prospects for long-term adaptability and survival. All the white abalone broodstock derived from an exceedingly small gene pool—the two cohorts that McCormick had raised plus the one very old remaining male retrieved from the wild during submarine dives back in 2000. In 2013, with a growing sense of urgency, the Bodega Marine Lab research team applied for a special Endangered Species Act (ESA) permit to collect more of the endangered white abalones from the wild, hoping to introduce new DNA into the aging captive population.[4]

With only limited captive breeding success during the first fifteen years of the program, the National Marine Fisheries Service (NMFS) had been reluctant to bring in more white abalone. Cautious biologists there and in other agencies thought that leaving some wild mollusks to reproduce on their own could be an effective second-basket recovery strategy to balance the risks of captive propagation. NMFS had continued to monitor the remaining wild deepwater white abalone population with follow-up ROV surveys in 2004, 2010, and in 2016 to track whether there was any natural recruitment. Surveys at San Clemente Island and at the remote reefs where white abalone were known to exist found steeply declining numbers and no new recruits at all, casting grave doubt on the possibility that the animals could ever rebound on their own.[5] Those grim findings galvanized greater agreement that a single wild white abalone could contribute more to the future of the species if it was brought into the captive breeding effort.

But meanwhile, there were some remarkable discoveries. Scuba divers with an Occidental College marine monitoring program found a single white abalone off the mainland coast in reef habitat that had been presumed vacant for decades! Coincidentally, around the same time, NOAA Fisheries Restoration Center biologist Dr. David Witting and his friend, engineer Bill Hagey, had started an informal project to look for remnant populations of white abalone; they also discovered some solitary animals.

How these few white abalones had turned up after such a long absence remains a mystery, but it inspired Witting, Hagey, and other friends to spend their weekends searching submerged reefs off the Palos Verdes Peninsula and Point Loma. White abalone have a cream-colored epipodial fringe but are usually heavily camouflaged with a leafy covering of red algae, so it was a "needle in a haystack" situation, says Witting. However, finding a few animals made the search compelling, and soon the small group of friends—mostly scientists—became a larger group of capable volunteers, intent on learning more. Methodically surveying larger areas, they found more individuals—though in total, only a skeletal population of mostly isolated animals, still too few to recover the population.[6] Nevertheless, discovering white abalone in shallower habitat off the mainland made research and recovery efforts more workable.

What started as informal surveys turned, over time, into substantial field expertise and useful data that helped convince federal regulators to finally issue a special ESA permit to allow collection of up to thirty wild white abalones for the captive breeding program in 2016. Witting and other divers from NOAA and the California Department of Fish and Wildlife (CDFW; finally renamed from "Fish and Game" in 2013 to reflect a more up-to-date sensibility about wild animals) set out immediately to retrieve them, but the rare animals proved extremely difficult to locate. By the end of 2016, the divers had been able to collect only a single male. However, over the next couple of years, Witting and other scientific divers identified preferred habitat and perfected methods of tracking and safe collection, ultimately bringing eleven more singleton white abalones into the captive breeding program. It turned out that most of the mollusks harbored the withering syndrome bacterium to varying degrees, but none showed symptoms of disease, underscoring both the prevalence of the pathogen in southern California waters but also these animals' apparent capacity to withstand it.[7]

In early 2017, Witting found a white abalone sitting on a rock not much bigger than its body so he brought the whole rock to the surface, minimizing stress and potential injury to the animal. Later named Green 312, the wild female abalone started to spawn millions of eggs as soon as she got to the surface, but the dive crew had no sperm to mix with her eggs. Kristin Aquilino drove down to southern California to ferry her back to Bodega Marine Lab and, in March, induced her to spawn with other white abalones, this time successfully mixing her eggs with gametes from a captive male. Aquilino gushed in email to colleagues: "Pretty cool that there are new genes in the program for the first time in 14 years!"[8] After the 2017 spawning season, sixteen thousand juvenile white abalones survived, a higher rate than ever before. The white abalone recovery program had crossed a successful but sober milestone: there were now more animals living in captivity than in the wild.[9]

Meanwhile, members of the White Abalone Recovery Team had realized that they needed more capacity—more people, more facilities, more research, more funding, and greater awareness—to succeed. Recognizing the value of broader collaboration, they pulled together additional partnering agencies and organizations—including universities, commercial abalone farms, nonprofits, and aquaria—to formally establish a new White Abalone Restoration Consortium. With an updated action plan, they divvied up responsibilities and projects—from research to outreach. With growing momentum, recovery coordinator Dr. Melissa Neuman and other NOAA scientists from southern California to Washington, DC, had advocated within their agency for more financial support, and in 2015, NOAA Fisheries finally gave white abalone

special priority through its "species in the spotlight" initiative, providing substantial additional resources and federal funding over the course of several
years to boost the ambitious recovery project.[10]

The new capacity came just in time. In spring 2019, when the Bodega
Marine Lab crew set about breeding the recently collected wild abalones, Green
312 showed her reproductive prowess once again, spawning an astonishing 20.5
million eggs in a single day. Another recently collected wild female and male,
plus well-conditioned captive animals, also spawned, producing millions more
fertilized eggs than ever before. By the end of the day, the BML crew had 19 million white abalone embryos, but the lab had capacity to settle only 1.5 million.[11]

With Neuman's leadership, the Restoration Consortium shifted into
high gear. They had less than a week to find homes for the larval-phase abalone
before the microscopic animals needed to settle. Aquilino overnight-shipped
two million embryos to the NOAA's Southwest Fisheries Science Center in La
Jolla, and 750,000 to Cabrillo Aquarium in San Pedro, and larvae were spread
out to several other facilities as well. At the same time, Aquilino arranged to ship
specialized food (diatoms and red algae) and other supplies needed to nurture
the baby abs. A week later, colleagues at ten different partnering aquaria, labs,
and aquaculture facilities housed larval white abalone in whatever containers
they could muster, and ultimately settled an estimated four million larvae. It
was an impressive show of support for white abalone restoration.[12]

However, beyond breeding enough animals, significant hurdles remain for restoring captive-raised juvenile white abalone into rugged ocean
habitats—in particular, predation, disease, and climate change. The team of
researchers in southern California at NOAA and CDFW, plus project partners
Paua Marine Research Group and The Bay Foundation, have taken the lead
on these fronts, using juvenile red and green abalone to evaluate various new
methods for outplanting—testing different protective structures, sites, and
timing for conditions with the coldest water, least predation, and greatest likelihood of survival. Hagey and Witting have also developed a system of remote
cameras for observing animal behavior, which promises to add important
clues that have long been missing about movement, feeding, and mating.[13]

Though withering syndrome—endemic in all waters of southern California—remains a challenge, the recent discovery of white abalone that can
withstand the disease has changed biologists' thinking. Although antibiotics
can give young captive-raised white abalone a grace period to become established in cold-water refuges that will help protect them, researchers are now
also conducting experiments to pre-expose juveniles to the bacterium ahead
of time. According to CDFW pathologist Jim Moore, disease resistance with
invertebrates occurs on a population level; typically, the 10 percent of animals

that survive are resistant. Pre-exposing the abalone knocks out the most susceptible in short order, in effect, hastening natural selection for the most resistant animals.[14]

Moreover, with growing concerns about rising ocean temperatures and ocean acidification, scientists at NOAA Southwest Fisheries Science Center are also conducting genetic analysis to determine which animals might be best suited to withstand these conditions—and diseases—into the future, with an eye to making certain that genes from the most resilient individuals become part of the recovering white abalone population. Tracking genetics of outplanted animals will be a critical part of the program moving forward.[15]

The white abalone recovery effort has found inspiration, too, in a parallel effort in the state of Washington to recover the closely related and severely depleted pinto abalone (*Haliotis kamtschatkana*), just recently listed in 2019 as a state endangered species. The Washington Department of Fish and Wildlife and the Puget Sound Restoration Fund, partnering with NOAA, universities, tribes, and other organizations in the region, began a captive propagation program in 2007, aiming to avert a California-style collapse of their local abalone populations. Biologists collected singleton pintos for broodstock, raised juveniles, and outplanted them every three years at ten sites in the San Juan Islands. After about eight years, they've been able to establish persistent aggregations at six reefs—though it's still too soon to know whether the outplants are successfully reproducing. The Washington researchers have realized it takes many years for pinto abalone to grow to a size that can readily be observed.[16]

Recovery coordinator Melissa Neuman says that restoration efforts with pintos help show what success will look like for white abalone recovery in California: "It will be an ongoing experiment; it's going to take time to figure out what works and what doesn't."[17] Neuman is optimistic about the first trial outplanting of captive-bred whites, which occurred in fall 2019, but she knows, too, that realistically, it will take many years—and favorable ocean conditions lining up just right—to recover reproducing populations. Ultimately, the White Abalone Restoration Consortium aims to propagate large numbers of juveniles, with as much genetic diversity as possible, and then to plant them out into the ocean, where, hopefully, the mollusks will find favorable food conditions and can draw on their inherent fecundity to start rebuilding populations.[18]

To follow efforts to recover the endangered white abalone over more than two decades underscores the difficulty of bringing back an endangered species, but the tenacity and passion of so many scientists and citizens involved is inspiring. As Kristin Aquilino explained, "We're in a make-or-break situation right now with the wild population. We put them into this mess, so it's up to use to save them."[19] David Witting thinks of white abalone recovery

14.1. In November 2019, scientists from NOAA Fisheries, the California Department of Fish and Wildlife, and many partnering organizations outplanted the first cohort of captive-raised white abalone in a historic effort to restore these endangered animals to southern California's marine ecosystems. (Photo by Adam Obaza)

in larger ecological and social contexts, too. By grazing on reefs, abalone keep open areas for settlement not only of their own offspring but also larvae of other invertebrates and diverse algae. "When we restore abalone," Witting says, "we're making the kelp forest habitat whole again." In the past, people came together around hunting and eating abalone; now, Witting sees hope in so many people coming together around the goal of restoring abalone.[20]

COMEBACK ROUND FOR BLACKS

While the intensive captive propagation effort is the best apparent hope for white abalone recovery, the most promising front for the intertidal black abalone lies on the remote and rocky coast of San Nicolas Island—the south-westernmost of the Channel Islands, sitting sixty-one miles offshore—where a small cluster of animals appears to be thriving.

These far-flung black abalone caught the eye of University of Washington–US Geological Survey researcher Glenn VanBlaricom. A benthic ecologist by training, Dr. VanBlaricom had first arrived on the island in 1979 to set up baseline monitoring plots as part of a sea otter translocation and recovery effort.

Because the remote San Nicolas Island falls under the jurisdiction of the US Navy and saw relatively little commercial fishing in its waters, the US Fish and Wildlife Service chose it as the place to translocate a group of sea otters in

1987. The federal wildlife agency's aim was to start a second population of the threatened marine mammals to avoid the potentially devastating peril of an oil spill in shipping channels closer to the mainland—an idea first hatched back in the 1970s, when Friends of the Sea Otter had battled with abalone divers and acceded to the compromise of moving some of the furry animals away from the Central Coast. Though commercial divers wanted to move sea otters out of state and opposed the translocation, the federal recovery plan ultimately chose San Nicolas as the best otter relocation site.[21]

VanBlaricom was tasked with tracking the sea otters' impacts in the intertidal zone, so he paid close attention to the black abalone that still crowded the rocky shorelines. As withering syndrome devastated the black abalone on all the other southern California islands through the late 1980s and early 1990s, the far-removed San Nicolas remained the only refuge where healthy black abs persisted in large numbers. Then, in June 1992, the disease hit hard. VanBlaricom witnessed the same pattern of decimation that had occurred everywhere else. By 2001, *Haliotis cracherodii* populations at San Nicolas had plummeted by more than 99 percent.[22]

However, the following year, VanBlaricom and his students noticed a new cluster of small black abs nestled into protected crevices at one of their intertidal monitoring sites. Apparently, some few surviving adults had managed to reproduce before they succumbed to withering syndrome during the massive 1997–1998 El Niño. Their young progeny grew, and each year, more juveniles appeared. It was one optimistic signal in a sea of bad news for black abalone.

At first, VanBlaricom suspected that the young abalones' survival was a fluke of local oceanographic conditions that delivered cold upwelled water to the particular site, offsetting the virulence of the pathogen. But as the animals endured, he couldn't help but wonder, might this be the pocket of disease-resistant black abalone that biologists had long been hoping for?

VanBlaricom ferried some of the mollusks back to the University of Washington's shellfish pathology lab for testing. (This was before black abalone was listed as an endangered species in 2009.) There, shellfish disease expert Carolyn Friedman and her graduate student Lisa Crosson designed an experiment to test whether the animals had developed resistance to withering syndrome. They exposed two sets of black abalones to the disease. One "naïve" set from near Carmel had presumably never before been exposed to the pathogen (the disease hadn't yet made it that far up the coast); the second set, from San Nicolas Island, had likely been fully exposed to the disease. Friedman and Crosson found that all the black abalone readily became infected; but, while the Carmel animals became sick, withered, and died, the San Nicolas animals seemed to withstand the disease, retaining their digestive gland function and their capacity to grip tight to rocks. The San Nicolas abalone seemed to have some trait that countered the pathogen's deadly proliferation in their guts. Examining DNA sequences, Crosson detected a more robust expression of genes related to immune response in the San Nicolas group. "After a decade of disease pressure," she explained, "the animals are definitely in the process of adapting."[23]

In the meantime, while testing red abalone from a Central Coast abalone farm, Friedman and Crosson observed what appeared to be an entirely new kind of infection: instead of turning red—the withering bacterium's diagnostic stain color—under their microscope, it turned navy blue.[24] Shifting to a higher-powered transmission electron microscope, they determined that the new infection was actually the same withering syndrome pathogen, but infested by a parasitic virus—a bacteriophage (derived from the Greek *phagein*, which means to devour). The bacteriophage appeared to attack the withering syndrome pathogen, rendering it less harmful to its red abalone hosts. Remarkably, the abalone farm reported fewer disease-related losses after the appearance of the phage virus.

Friedman and Crosson wondered whether the bacteriophage might have the ability to knock back withering syndrome in the black abalones they were testing for disease resistance. They did a second experiment with the naïve Carmel and disease-selected San Nicolas Island black abs. The results were stunning. The abalone exposed to the phage-infested pathogen—animals from both Carmel and San Nicolas—endured the infection much better and with fewer symptoms; their mortality appeared to be delayed and significantly reduced compared with the non-phage-infested animals.[25]

Beyond potential disease resistance, the trial results showed that the black abalone might benefit from the unexpected and remarkable ecological appearance of the bacteriophage, which also proliferates during extended periods of elevated ocean temperatures. In the past few years, the researchers have found that the phage has now spread through most of California's marine waters where the withering syndrome pathogen is present.[26] The emergence of the bacteriophage suggests a greater level of complexity in marine ecosystems than what fisheries managers had ever presumed. Considering pathogens and parasites as ecological players, researchers now aim to understand differing virulence of microorganisms and susceptibilities of hosts—all dependent on variable sea temperatures and ever-evolving genetics. Understanding the potential risks of disease under shifting environmental conditions will be crucial to managing and restoring abalone in the face of climate change.[27]

However, the discovery of the new and entirely unanticipated bacteriophage has kindled a new hope for the natural recovery not only of black abalone but of all vulnerable California abalone populations. With the bacteriophage now spread throughout the range of withering syndrome–infected waters, it may give the stressed and imperiled mollusks a fighting chance to survive and endure. The resilience of nature just might prevail.

* * *

Back on San Nicolas Island, that small cluster of healthy black abalone continues to thrive but could face another challenge. Glenn VanBlaricom's initial job had been to monitor intertidal invertebrates before the arrival of sea otters translocated in 1987. After their release, the majority of otters quickly dispersed and most died, leading many to question the success of the effort.[28] However, according to counts in 2017, the progeny of that small number of translocated sea otters had finally—thirty years later—begun to surge to nearly 130 animals.[29]

What would happen if the "threatened" sea otters start eating this disease-resistant population of "endangered" black abalone? VanBlaricom takes a broader perspective, explaining, "The end game is to restore a functioning ecosystem, not just one animal instead of another." He hopes that the black abalone will soon become well established enough to withstand some otter predation by hunkering in the safety of crevices as they have for millennia. He points to a 2015 study of black abalone and sea otters on the Central Coast—forty years after otter recolonization—that showed greater numbers of abalone in areas with otters than without, likely owing to the more-ample drift kelp supply for mollusks tucked into protective, cryptic habitat since otters keep kelp-eating urchins in check.[30]

In the lineup of existential threats, VanBlaricom and other marine ecologists involved in recovery efforts regard elevated seawater temperature, ocean acidification, poaching, and degradation of habitat as far graver hazards to the long-term survival of black abalone than are the otters.[31] Nevertheless, they plan to keep careful watch on vulnerable populations of both sea otters and black abalones—two beloved and imperiled California animals, predator and prey, persisting together in one of southern California's few remote outposts.

* * *

Meanwhile, there's also been a comeback of black abalone at several spots in Channel Islands National Park. When I first started this book, I had the privilege of going out to San Miguel Island with National Park Service biologists Dan Richards and Steve Whitaker to see the very small remnant population of black abalone there.[32] Richards, who started his career as part of the crew that

14.2. After nearly three decades, small populations of what appear to be disease-resistant black abalone are finally starting to make a tenuous comeback at remote intertidal reefs on the Channel Islands. (Photo by author)

set up the park's initial monitoring plots in the 1980s, had watched firsthand as the abundant intertidal black abalone populations withered and crashed, and then painstakingly continued to observe the Channel Islands' intertidal habitat, looking for any sign of the animals.

Whitaker was new to the job in 2009, but over the past decade has been in a position to track the tenuous beginnings of black abalone's natural recovery. He first started to see mollusks return on Santa Cruz Island, where populations had dropped to just one but then rose into the hundreds. Not until 2016 did he begin to find growing numbers of juvenile animals on San Miguel Island. The slow but hopeful rebound reveals that nature's time frames may be very different than our own, which is one reason Whitaker says collecting long-term data is so important.[33]

In the face of what look to be many insurmountable challenges and perilous close calls, I remind myself that California's abalone have endured for more than seventy million years. Black abalone may well have what it takes to survive—if only we can allow them space and time.

COMEBACK ROUND FOR GREENS

While the National Marine Fisheries Service focused on averting the extinction of white and black abalones, and the California Department of Fish and Game focused on managing red abalone in northern California's sportfishery and considering the proposal for a commercial red ab fishery at San Miguel Island, California's other abalone species were mostly on their own. Recognizing a responsibility to prevent future listings and extinctions, NMFS had in 2004 designated green, pink, and pinto abalone as federal "species of concern"—a designation for animals likely headed toward threatened status that needed monitoring and support. However, owing to lack of funds and limited staff at both federal and state levels, these less-imperiled species garnered little attention.

Yet, of all the species, southern California's green abalone seemed to be best suited to cope with the changing conditions at hand. They had a natural capacity to thrive in shallower, warmer waters and seemed to coexist with withering syndrome better than the others. For these reasons, abalone researchers, including Mia Tegner, had proposed that greens might be the most promising species for restoration efforts in southern California, but the idea sat simmering on the backburner—until Nancy Caruso arrived on the scene.[34]

Caruso began her marine biology career in aquaculture in Florida, but she was also an avid recreational diver, attentive to the natural dynamics of the undersea world. When she moved to southern California and started to meet other local divers, she heard incredible stories about the good old days, when abundant green and pink abalones studded the reefs and divers gathered great numbers of the succulent shellfish for beach parties. Soon she became involved in efforts to restore kelp forests off Laguna Beach, ultimately taking the lead in training hundreds of volunteer divers. Meanwhile, she found her passion as a marine science educator, bringing kelp restoration projects into thirty local schools. She helped kids grow kelp in tanks in their classrooms; then she and her volunteer divers would plant it out on depleted reefs. Nancy also spearheaded the popular Laguna Beach Kelp Fest, pulling in her volunteers, teachers, and kids to participate, inspiring thousands of people with the vision of restoring southern California's dwindling kelp forests. When the kelp began to thrive, Nancy was eager to take her community-based restoration vision to the next level. She wanted to restore the kelp forest's traditional denizens, and green abalone was at the top of her list.

When Nancy taught school kids about animals of the kelp forest, she realized that none of them even knew about abalone. She gave them the assignment to ask their grandparents, and they came back with stories about pretty shells, how good the abalone tasted, and how many there had once been. Struck hard by what had been lost, both biologically and culturally, Caruso decided she had to do something. She began in her bedroom, growing green abalone in a small aquaculture tank as a pilot project. Then in 2010 she introduced tanks into seven classrooms, where she helped kids raise baby green abalones. It was

Nancy's dream for kids to raise the mollusks and then put them out into the wild. She hoped it would it help to restore reefs but also excite the kids and give them a sense of pride and concern for nature. "I want to inspire the next generation of marine biologists and stewards for the future," she told me.[35]

I met up with Nancy that first year she had school kids growing green abalone. She took me to classrooms where kids had already become ardent abalone enthusiasts. At Warner Middle School in an Orange County neighborhood known as Little Saigon, seventh graders told me that, before their project, they'd not really known about abalone. One girl had heard about abalone only as food—not as an animal. When we gathered around the aquarium tank that held six mollusks the size of quarters, the students explained to me how they took turns tending water quality, feeding strands of kelp, and measuring growth of the animals' vivid green shells each day.[36]

Not only did Nancy engage kids in their classrooms but she also arranged for them to teach others what they had learned. Kids designed camo-patterned Abalone Army T-shirts, and at the Aquarium of the Pacific in Long Beach, they set up a display and told thousands of visitors about their project, spreading the hopeful vision of restoring the iconic shellfish to southern California's waters.

However, even with careful attention to sanitation, some of the classroom-raised abalone got sick with a mysterious ailment. "It was a good learning experience for the kids," Nancy later told me, "but not a good way to raise healthy animals for outplanting."[37] With high concern about pathogens spreading to vulnerable remaining wild abalone, the CDFG had issued no permits for outplanting since the sabellid infestation of 1996.[38] So when Nancy proposed to outplant her classroom-raised green abalone, CDFG abalone biologist Ian Taniguchi leveled with her, "If you want to persist with this project, it's going to be hard"—which, she says, only kindled her stubborn resolve.[39]

When Caruso applied for a permit to outplant green abalone, the CDFG convened a team of expert advisers to consider concerns about disease and genetics and conditions to be met before her project could proceed. Since green abalone were not yet listed as endangered, the biologists realized that the animals might serve as experimental proxies to help develop better outplanting techniques to restore the more imperiled *Haliotis* species. [40]

Concerns about disease risk were addressed when pathologists determined that all wild green abalone in southern California were already infected with the ubiquitous withering syndrome bacterium, though they didn't show symptoms. Concern about nonnative shell-boring sabellid worms was quelled when all the state's aquaculture facilities were officially certified as sabellid-free.[41] Concern about genetics required additional study, but, ultimately, researchers found little genetic difference between green abalone on the islands and mainland, indicating that greens could be outplanted without risk of diminishing genetic variability.[42]

By the time Nancy finally got a state permit to outplant green abalone in 2012, she'd realized that her classroom-grown animals couldn't meet the bar of being viable in the wild. But when she went to buy more juveniles, she found that aquaculture farms had all but stopped growing green abalone since market demand was highest for reds. The Cultured Abalone Farm in Goleta offered to sell her all its remaining greens—sixty-seven adults, each about 5 to 6 inches long—for broodstock.

Although the genetic study had shown that all wild green abalone shared a near-identical genetic profile, it turned out the farm-raised greens were different. Nevertheless, with Nancy pushing hard, the CDFG issued her a limited experimental permit.[43] She could outplant the farm-raised green abalone at Crystal Cove State Park near Laguna, if she retrieved them all in two years. It would be a trial run to see whether larger-sized abalone could be successfully outplanted and monitored.

Caruso rose to the challenge. With more than a decade of social capital invested in the vision of restoration (Nancy had told me, "My entire social life is based on restoration"), she drew on students to create special tags for the abalone and then marshaled her dive volunteers to monitor the outplanted animals every forty-eight hours. At one point, the volunteer patrol crew found a mass of hammer-crushed shells—evidence of poachers—but after Nancy installed signs warning of continuous surveillance, the problem stopped.

After two years, in the fall of 2014, Caruso retrieved all the outplanted green abalone to meet her permit requirements. She determined that their survival rate was 40 percent, much higher than in any previous outplanting effort, almost certainly because she used large adults rather than juveniles and, of course, because her dedicated volunteers had doggedly watched the site.[44]

Caruso now wants to build on her success with an even more ambitious plan to restore green abalone. Her hope is to help create several pockets of adults that can spawn enough times to eventually repopulate Orange County's still-empty-of-abalone nearshore reefs. And she hasn't given up on the idea of raising some of the abalone in classrooms to inspire local kids. So far, with a new state permit, Nancy has collected wild green abalone broodstock, and she's arranged for the Cultured Abalone Farm to spawn them and raise the juveniles for a year. Then she will transfer the yearling abs to local aquaria and classrooms to be tended by students for five more years—until they're large enough to better withstand predation.[45]

As much as Caruso aims to reverse the long decline of green abalone, she aspires, too, to make sure there will be a community of up-and-coming southern Californians to care about abalone's return—and to steward the kelp forest ecosystems that will be needed to sustain them.[46]

Meanwhile, after a more than twenty-year break from fishing pressures, and with enhanced patrol efforts around southern California's Marine Protected Areas, divers have begun to notice the hopeful reappearance of small numbers of wild green abalone in some places.[47] If ocean conditions cooperate, with several years of successful recruitment and abundant kelp for food, could the wild green abalone rebound on their own? That remains to be seen.

CALM BEFORE THE STORM

Along California's wave-blasted shoreline, the Mendocino coast had long been a northern stronghold where the state's greatest cache of abalone endured. Every year thousands of sport divers came to ply the cold, kelp-entangled waters for their chance to pop a red abalone—to experience the foraging that has been a vital California tradition since people first showed up more than ten thousand years ago. With the commercial fishery closed, the only way to eat an abalone was to hunt it yourself—or to be lucky enough to have an abalone-diving friend who was willing to share.

It was the summer of 2013. Sport season was open, highway pullouts were crammed with cars, and campgrounds were full. In a buoyant yellow sea kayak, I paddled seaward from Russian Gulch State Park. The water lay calm, but just offshore, beyond a protective lineup of jagged rocks, I could see the chaotic lurch of waves. I was following Dan Richards, the National Park Service biologist I'd accompanied on a search for rare, endangered black abalone on San Miguel Island several years earlier. We'd kept in touch. Like so many Californians, Dan made a pilgrimage up to the North Coast each summer to camp in the redwoods and dive for red abalone. He'd been coming for decades, and this year invited me to join him to experience the other end of the abalone spectrum—the abundant part, where the animal became a delicious food.

In our kayaks, we rounded a cluster of rocks at the mouth of the cove and paddled south to another protected spot. Dan tied our boats to a bull kelp, pulled on a neoprene jacket over his old farmer-john wetsuit, buckled his weight belt, tightened his mask, put a snorkel in his mouth, slipped into fins, grabbed his pry bar, took one deep breath, then another, then dropped into the clear seawater. I donned my own gear and followed him.

Colorful seaweeds waved back and forth in the undulant water. I watched Dan's blue fins deep below me. He was scouting the submerged rocks for abalone big enough to feed everyone back at camp. When we ascended to the surface for a breath, he invited me on a tour. We submerged, this time swimming side by side, and he silently pointed out different seaweeds, fish, crabs, urchins, and, of course, the ovoid shells of red abalone clinging to rocks, under camouflage cover of algae, sponges, barnacles, and everything else that grows on their backs in the pile-on lifestyle of sessile marine animals. Once my eyes fixed on their shape and size, I saw red abalone everywhere.

Before long, with the ocean's cold penetrating our thin wetsuits, Dan was ready to take his quarry. Following down as best I could, I watched him locate a big ab, measure it, pop it quickly, and then press it to his belly, where the animal quickly took hold on his wetsuit as we surfaced again for a breath. At the time, state regulations allowed Dan to take two per day, so he kicked straight down, deeper than I could manage, and in short order returned with another large abalone stuck to his belly. We swam back to the boats, where he put the animals into a mesh sack. Up and down with the swells, I began to feel queasy. Dan made abalone diving look easy, but even on a calm day like this, it required skill, experience, knowledge, and a certain equanimity in the ocean. One of the unspoken satisfactions of abalone was that they had to be earned.

* * *

Back in camp, redwoods towered all around, with late-day sunlight filtering through leafy foliage. After warm showers and visits with other friends joining us from southern California, we were ready for the next "ab camp" ritual. Already, the air was filled with knocking sounds, as if a flock of woodpeckers had descended into the forest. It was campers pounding abalone. Some pounded on the posts that demarcated campsites, others pounded on big old stumps, and still others brought special wood blocks from home.

Dan retrieved the red abalones from his mesh sack, and we took a closer look at them. They were still alive and still seeking to grip to any protective surface. One clung to my hand, sticking tight to my fingers with its thick foot. When I finally pulled the animal off, I could see the impression of my hand in its flesh, kind of like memory foam. Ever the biologist, Dan showed me the animal's molluscan anatomy: its tentacled epipodial fringe, its gonad, its gills, its eyes.

Biology aside, it was time for meat. In one quick move, Dan shucked the living animal from its shell and began to trim off its tentacles until he had a hefty puck of white flesh. Many traditionalists would have sliced it into steaks for pounding, but Dan wrapped it in a dishtowel, dug out a small wooden bat, and pounded the ab whole. The taps of his pounding joined the rhythm of the greater campground. After a while, he passed the bat. When it was my turn, I couldn't help but launch into the one verse I remembered from the traditional song:

We sit around,
And pound and pound,
But not with acrimony,
Because our object is a gob,
Of sizzling abalone.

The congenial group laughed. When Dan deemed the two abalone pucks sufficiently pounded, he took out a slender knife and proceeded to filet the meat into thin strips. When he was a kid, his family always ate abalone fried in cracker crumbs with a side of baked beans. This time, Dan's wife Sara put a few thin strips into a small bowl of fresh-squeezed lime juice to make a ceviche appetizer. Then she fired up a Coleman stove on the picnic table, heated a skillet with olive oil, butter, and chopped garlic, and fried the abalone strips until lightly crisped. We all lined up with blue-speckled camp plates to sample the meat, with garlic-buttered French bread and salad on the side.

It's a modern habit to wolf down foods with absolutely no awareness of provenance, but after joining Dan to hunt and pound the abs—my head swimming with deep history—I realized I'd never before eaten a bite so laden with story. To eat a wild abalone in California was a unique privilege. I savored it, letting the meat fill my mouth, and then tried to find words to capture it: delicious, savory . . . something like scallops, salty . . . like its mother sea. I remembered the words of abalone lovers I'd talked with over the past several years. A man who grew up diving in Malibu told me eating raw abalone was like "eating pure energy." A woman who gathered abs on the North Coast as a child told me it was "communion." I could now attest that the experience of eating abalone was about far more than flavor.

As I ate the abalone strips in the company of friends and regal redwoods, I cherished the hallowed tradition of abalone foraging and the remarkable animal that made the day of vigor and fellowship possible. With empty shells glimmering in the campfire light, I marveled again at their reflective qualities, remembering how ancient peoples all around the Pacific Rim associated abalone shell with vision—both literal and figurative—and wondered what essential truth the shell might be showing us now.

That was early summer of 2013, which now on the North Coast seems like a lifetime ago. None of us in the campground that night had any idea what would happen next.

PERFECT STORM FOR REDS

Northern California's red abalone had long been considered the state's most robust *Haliotis* population, protected by tight fishing restrictions that banned scuba and commercial take, leaving a bank of deepwater broodstock that continually replenished the nearshore supply for avid free divers and shore-pickers who returned year after year.[48] Nevertheless, starting around 2000, California Department of Fish and Game (CDFG) biologists became aware of a prolonged period of depressed recruitment of young red abalone and worried that continued heavy fishing on non-reproducing populations could threaten the long-term viability of the popular recreational fishery. They'd learned this lesson in the wake of El Niños in the south and wanted to apply it proactively in the north. In response to their recommendations, in 2002, the California Fish and Game Commission cut each sport diver's annual take of abalone from one hundred animals to twenty-four.[49] However, the reduction in fishing pressure did not lead to the hoped-for population rebound. Instead, through subsequent surveys, CDFG divers identified a troubling trend of declining red abalone densities.[50]

That was the backdrop when a toxic red algae bloom hit the Sonoma coast in the summer of 2011. Within a short period, thousands of poisoned shellfish and other invertebrates washed ashore dead. At the popular Fort Ross dive site, 30 percent of abalone were killed. It was the largest marine invertebrate die-off ever recorded in the region.[51]

No one realized it at the time, but the red-tide die-off would be the first in a series of ecological shockwaves to hit northern California's nearshore marine ecosystems, putting red abalone at heightened risk. Two years later, in 2013, a grisly epidemic wasting disease started to attack sea stars up and down the West Coast from Mexico to Alaska, causing them to lose legs, deteriorate, and die off en masse. The near-total loss of these important sea urchin predators set off a coast-wide explosion of purple urchins. In northern California, the small urchins grazed voraciously, turning lush kelp forests into "urchin barrens" reminiscent of those that had plagued southern California back in the 1970s, but far more extensive. Within a few years, purple urchin populations reached densities estimated to be more than sixty times greater than at any time in recorded history.[52]

Then, northern California's marine ecosystems were buffeted by historically unprecedented incursions of warm water. In 2014, warm seawater masses from the North Pacific's so-called Blob dipped down into northern California. The following year, a "Godzilla" El Niño—the most severe on record—pressed warm waters northward, adding to the thermal stress and amplifying the threat for northern kelps, since warmer water holds fewer nutrients they need to grow.[53] Because northern California's dominant canopy- and habitat-forming kelp—bull kelp (*Nereocystis luetkeana*)—are annual plants, growing up from spores each year (unlike southern California's perennial giant kelp, *Macrocystis*

14.3. Dr. Laura Rogers-Bennett started the state's long-term monitoring of northern California's kelp forests, including studies of red abalone recruitment. Here she orchestrates scientific divers on a research cruise. (Photo by author)

pyrifera), they cannot readily grow back without successful reproduction. Moreover, young spore starts and tender juveniles are particularly vulnerable to urchin grazing. Aerial surveys soon revealed that the North Coast's kelp forests had been reduced by 93 percent.[54]

With each ensuing calamity, the California Department of Fish and Wildlife's biologists grew more concerned about the capacity of North Coast abalone populations to endure mounting environmental stresses while also sustaining the ongoing recreational fishery—with more than 225,000 animals taken each year, not including those lost to the persistent scourge of illicit poaching.[55] Leaning heavily on guidance from the 2005 Abalone Recovery and Management Plan (ARMP), they conducted dive surveys at eight index sites to track red abalone density and recruitment success—the precautionary criteria intended to trigger specific fishing closures, shortened seasons, and harvest cutbacks to conserve sufficient broodstock to maintain the fishery.

In 2013, when department surveys confirmed lack of recruitment and that red abalone density had dropped below a critical ARMP threshold, biologists recommended that the state Fish and Game Commission reduce the total allowable catch by 25 percent, as specified in the plan.[56] North Coast shore-pickers and sport divers had a long tradition of conserving North Coast abalone and had strongly supported development of the ARMP, but this time a new generation of divers questioned the need for further restrictions. New leaders at the long-standing Sonoma Coast Abalone Network (SCAN) and at The Watermen's Alliance, a new group formed to represent divers in the controversial public process to designate Marine Protected Areas on the North Coast, supported a later starting time and site-specific closures but disagreed with an across-the-board lowering of bag limits. Recreational divers' main objection was that the department's ARMP survey protocol relied on too few index sites, including some especially high-use areas, which they thought did not represent the health of the whole abalone population. (The ARMP rationale was that high-use sites would best serve as an early warning to detect too-high fishing pressure.) Some sport divers argued that the recent Marine Protected Area closures were already sufficient to protect abalone broodstock. Others contended that more fishing restrictions punished honest divers while giving poachers a pass.[57]

Although sports divers objected, the commissioners decided to act in accordance with the ARMP guidance. They cut the total catch for the 2014 season by ratcheting back annual bag limits from twenty-four to eighteen red abalones per diver (nine in Sonoma) and setting a later start time to eliminate high-impact, early morning, low-tide shore-picking.[58]

In the following years, more than twenty-five thousand avid recreational divers and shore-pickers bought fishing licenses and abalone "report cards" and returned to the North Coast to hunt for red abalone.[59] Meanwhile, the waters of the North Pacific had never before stayed so warm for so long. By the fall of 2015, with expanding purple urchin fronts and diminishing kelps, CDFW biologists found that abalone density at three index sites had dropped to zero.[60] That winter, department environmental scientist Dr. Cynthia Catton gave a sobering presentation to the state Fish and Game Commission, describing the alarming cascade of large-scale ecological stressors as a "perfect storm."[61]

When the spring 2016 abalone fishing season opened, more sport divers saw for themselves the grim decline of kelp. One longtime Calistoga diver described a favorite dive spot near Point Arena as being "like a desert." Many reported large numbers of dead abalone on rocks and beaches.[62] However, some divers found pockets where abalone and kelp persisted, making them hopeful that perhaps conditions were less dire than state regulators claimed.[63] Later surveys would document a confounding phenomenon—when hungry abalone moved from deep waters into shallow areas searching for food, it gave the mistaken impression that the population might be recovering.[64]

Through the summer and fall of 2016, as CDFW divers continued surveys to track abalone health and densities, they found skinny abalone with shrunken gonads, very low rates of juvenile recruitment, and severely limited kelp. Most significant, they found that abalone densities—even in deeper waters—had dropped below the next critical ARMP threshold, triggering the need to again cut back fishing pressure. The department alerted the Fish and Game Commission and, in December, proposed an emergency rule change to cut annual bag limits in half, from eighteen to nine abalones.[65] Again, sport divers pushed back but worked with the department to come up with a compromise: shortening the next year's fishing season by two months and reducing bag limits to twelve abalones per year.[66]

Despite wishful hopes, undersea conditions continued to deteriorate off the northern California coast. In spring 2017, CDFW divers reported 25 percent of abalone starving at nine index sites and with shrunken gonads for a second year in a row, clearly indicating poor prospects for reproductive success.[67] Later that year, follow-up surveys at eleven index sites—with an added crew of dozens of independent scientific divers from PISCO (Partnership for Interdisciplinary Studies of Coastal Oceans) and Reef Check—found still-poor environmental conditions, with ravenous urchins now scraping even coralline algae from rocks. Most alarming, they affirmed the continued lack of abalone in deeper waters, long considered critical as a protected nursery for the North Coast fishery's broodstock.[68]

That summer, with abalone densities fallen below the ARMP threshold for complete closure of the fishery, the department alerted the Fish and Game Commission that still more restrictive regulations would be needed. CDFW Invertebrate Program manager Sonke Mastrup warned that persistent starvation conditions for abalone and poor reproduction "add to a building conclusion that the current model of fishery is not sustainable."[69]

Ironically, even as environmental conditions worsened, the Department of Fish and Wildlife was in the process of developing a new fishery management plan (FMP) for red abalone, to update the ARMP. Through the department's participatory public process to develop this new FMP, recreational divers—echoing concerns raised by commercial divers who had engaged in the effort to open a fishery at San Miguel Island—raised questions about the CDFW's precautionary, density-based management, suspicious that it could prematurely or unnecessarily close access to abalone fishing.[70]

Starting in 2014, some recreational divers had joined together with The Nature Conservancy in an alternative Abalone Working Group to develop an entirely different approach to manage red abalone fisheries—using fishermen-generated landings data (lengths of abalone harvested) to model abalone populations with the aim of allowing for the maximum sustainable catch. They contended that working collaboratively with sport fishermen to obtain fisheries-based data, generated with a new mobile app, would be less expensive, more efficient, and also more responsive than the CDFW's fisheries-independent density approach, which relied on periodic snapshot index surveys by the state's scientific divers.[71]

Because the department's FMP process unfolded at the same time that environmental conditions on the North Coast deteriorated and plummeting abalone densities triggered a succession of increasingly restrictive fishing regulations to safeguard dwindling broodstock, the two issues were often discussed together, with longtime stakeholders rehashing long-contested points. With lingering animosity and skepticism about CDFW's precautionary management, some recreational divers put more stock into The Nature Conservancy's modeling approach, believing it would allow for continued and more abalone fishing. They pressed the commission to hold off on any further restrictions until a new collaborative fishery management plan based on The Nature Conservancy proposal could be adopted.[72]

In the fall of 2017, the commission considered updates from department staff and public comments from divers, and at its December meeting in San Diego was poised to make another gut-wrenching decision about the fate of abalone. This time, several sport diving groups and former commercial divers opposed a full closure on principle, fearing that northern California's abalone fishery would never again be reopened. While recognizing the dire circumstances, they urged commissioners to keep the fishery open, at even a low level. Joshua Russo, a North Coast guide and diver who had taken leadership as president of Sonoma Coast Abalone Network (SCAN) and also The Watermen's Alliance, expressed the sport divers' dilemma: "I don't want to see the fishery collapse, and I don't want to see the fishery go away."[73] Wayne Kotow of the California Coast Alliance said he understood the need to do something but expressed frustration that the Department of Fish and Wildlife only ever closed abalone fisheries and never opened them back up. "What is the plan for being proactive?" he asked, even as he argued against the agency's proactive recommendation to close the fishery early enough to shore up the base broodstock for the future.[74]

However, ocean conservation group leaders and some individual divers spoke in favor of following the ARMP and closing the fishery. Geoff Shester from the marine advocacy group Oceana warned, "This is a species that will go to a point of no return if we're not careful." North Coast diver Brandy Easter described that "it was extremely heartbreaking to dive in areas that were once thriving and abundant with life that are now kelp-less urchin barrens with atrophied abalone and empty shells." Worried that weakened abalone would not survive upcoming winter storms, she urged a temporary closure, adding, "The abalone fishery needs our stewardship not our selfishness."[75]

Twenty years had passed since an earlier Fish and Game Commission had wrestled with the anguishing decision to close the red abalone fishery, commercial and recreational, south of San Francisco—a history that some in the audience had lived through but that current commissioners seemed not to know. CDFW Invertebrate Program director Sonke Mastrup quickly recapped the whittling away of broodstock owing to otters, overfishing, and disease that had ultimately pushed southern California's abalone over the edge. "Unfortunately, the Commission at the time and other decision makers . . . pretty much waited until it was all collapsed," he explained. Underscoring that the North Coast's red abalone now faced similar danger, he recommended a complete closure of the fishery to allow for a better chance of recovery. "You cannot escape the biology of this animal," he continued. "It is not the most productive animal in the world particularly when you drive densities down to where you really impact the reproductive output. That's where we believe we are . . . at risk of jeopardizing future productivity."[76]

Still, the commission vacillated—torn between allegiance to sport divers, with their passionate drive to continue the North Coast's time-honored tradition of recreational abalone diving, and trust in the Department of Fish and Wildlife biologists, with their science-based recommendations and strong sense of responsibility to protect red abalone populations from the risk of collapse. If ever there was a time to draw on lessons from the past—and to err on the side of precaution—this was it.

Ultimately, the commission voted unanimously to follow the ARMP and close the North Coast's red abalone fishery—for a year. The next year, with persistently poor ocean conditions and dead animals washing ashore, commissioners would extend the closure to 2021.[77]

For more than a century, California had tried to manage its distinctive shellfish for sustainable fisheries. After precipitous declines nearly everywhere else, the North Coast's breath-hold sport fishery had long been regarded as a bright spot of abalone management success. But now, even the most precautionary fishery management plan could not safeguard abalone from looming new threats.

* * *

The following spring, 2018, when divers and shore-pickers would have typically flocked to the North Coast's emerald coves and rocky reefs in search of red abalone, dive club leader Joshua Russo worked to tap their passion for diving and turn it to the purpose of restoration. Calling for volunteer divers from throughout California to pull together, he organized several events to remove the still-burgeoning purple urchins from some of the North Coast's coves and reefs. Facing the staggering scope of urchin barrens—extending for hundreds of miles with no kelp in sight—scientists from the CDFW and the Greater Farallones National Marine Sanctuary had worked with kelp experts and partnering stakeholder groups to develop a strategic plan for kelp forest restoration. They set the goal of maintaining a set of protected coves where bull kelp can endure as spore stock for the future when favorable ocean conditions return.[78]

As part of this initiative, called KELPRR (Kelp Ecosystem and Landscape Partnership for Research on Resilience), Russo not only organized sport divers to remove purple urchins from shallower areas, he also helped raise hundreds of thousands of dollars from local counties and sport divers statewide to hire commercial urchin divers to bring their boats and hookah gear to clear out purple urchins in deeper areas.[79] Other community members are trying to find uses for hundreds of thousands of emaciated purple urchins, hoping to generate revenues for future urchin management. Volunteer divers, with the ocean monitoring organization Reef Check, are also helping with ongoing surveys of sites along the North Coast, adding to data being collected by the CDFW's long-term kelp forest monitoring project. According to Reef Check's Tristin McHugh, "The silver lining in all of this may be the way scientists and citizens are coming together to find solutions."[80]

With urchin barrens likely to persist and marine heat waves predicted to become more frequent and intense, finding ways to shore up the resilience of the North Coast's marine ecosystems will be crucial if red abalone and other animals that depend on local kelp forests are to survive into the future.[81] In this, the communities and coves of northern California will no doubt become a critical testing ground.

✳ ✳ ✳

In the meantime, following the direction of the Fish and Game Commission, the California Department of Fish and Wildlife continues its project to develop a new fishery management plan for red abalone, with ongoing participation from avid former commercial and sport divers, still wanting to hunt for abalone—if not in northern California, then perhaps someday in central or southern California.

If and when California's abalone ever do rebound, the critical question remains: Have we learned the lessons we will need to coexist with these animals into the future?

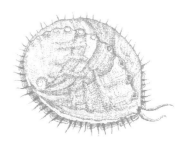

Conclusion

POISED AT THE BRINK

When I started this book, I was intrigued to learn about the history of a unique animal with an iridescent shell that was deeply embedded in the heritage and culture of communities up and down California's coast. Since then, I've listened to the song of shells in traditional Native American dances. I've been regaled with stories of memorable dive trips, family campouts, and beach parties. I've had the privilege of eating red abalone. I've shared a sense of outrage with people who fought bitter political battles to conserve abalone for the future. I've learned about the rigors of diving and became fascinated by the challenges of marine research. I've had the chance to bear witness to the beginnings of recovery for black and white abalone but also to the sobering collapse of northern California's kelp forests. Through it all, I've been impressed by people's abiding devotion to abalone and moved by the poignancy of what's been lost as these shellfish that once connected so many of us to the ocean have become imperiled.

The history of abalone shows how humans have disrupted and irrevocably changed marine ecosystems in ways that we are still coming to understand. After nineteenth-century colonizers misinterpreted the historical and ecological circumstances that gave rise to abalone abundance, Californians inherited a fond but false story of natural bounty that led to the flawed assumption that abalone was an inexhaustible seafood. After decades of heavy fishing eroded the unnatural surplus, the slow-growing animals became more vulnerable to still-rising fishing pressures and changing environmental conditions. Even as abalone stocks collapsed, politicized fisheries management made it impossible to nimbly address accreting threats that ultimately put the mollusks at risk.

Knowing the deeper history of our marine environments is important because the story we tell ourselves about how we arrived at the situation we face today creates the stepping-off place for the future. We must strive to find a common understanding based on history and science, but also on love and care, because the need to make consequential decisions about the fate of people, wildlife, and ecosystems far into the future will only become more pressing.

As we now enter a period where climate change is rocking the foundations of ocean ecosystems, the history of California's abalone points to important lessons we must heed if we want our precious marine life to persist into the future. Abalone have already been sentinel animals, plainly revealing the vulnerabilities of all marine life to changing ocean conditions—especially the increasingly frequent and severe El Niños and marine heat waves that have knocked out abalones' food, constrained reproduction, and created a hot bed for devastating disease. Similar threats have affected abalone and marine ecosystems around the world and underscore the urgent need for society to tackle the global climate crisis.

While the tragic decline of abalone illustrates the failure of the traditional management paradigm that California used to manage its fisheries for most of the twentieth century, the California Department of Fish and Game's development of the Abalone Recovery and Management Plan (ARMP) at the outset of the twentieth century represented a significant turnaround. Learning from past mistakes, the department shifted gears to base management on science, including the unique biology of abalone and data cued to conditions of stocks—aiming finally to take a precautionary approach instead of waiting until after a crisis to take action. As conditions in northern California's kelp forests deteriorated, the ARMP passed a critical test, giving decision makers data and guidance needed to proactively reduce fishing pressure on red abalone. However, the environmental nature of the predicament highlights the growing need for approaches that go beyond just managing fishing pressure on individual species. With predictions for more marine heat waves, changes in ocean chemistry, and rising risks for disease, far greater attention needs to be directed toward protecting and restoring resilience to marine ecosystems, too.

Looking ahead, to bring abalone back to cracks and crevices in our coastal rocks and reefs, we will need to carefully set the stage for restoration and natural recovery of each species. Many factors aligned to decimate California's abalone, and now we will need the right conditions to align to foster their return. As I reflect on this history, I am struck by the fact that scientists still grapple to understand what healthy natural populations of different abalone species might actually look like, since much of what is known about these animals derives from times of either unnatural abundance or utter scarcity.

For that reason, we need to think of restoration not as taking us back to some imagined previous state of abalone abundance but rather as taking us forward to ecosystems with diverse and resilient communities of plants and animals, where enough abalone have the genetic variation, proximity, and food they need to survive, even with rising ocean temperatures. Knowing that abalone thrive best with ample kelp, we must also reconcile the deeper history of our marine environments and come to better understand the role that sea otters and other urchin predators, such as seastars, have played in kelp forests in the past to consider whether we may need them to play such roles once again.

Although looming threats from climate change remain daunting, I find tremendous hope in reports from some divers that small numbers of pink and green abalone have begun to reappear in some spots in southern California reefs after twenty years of fishing closure. I am optimistic, too, that efforts to recover the most imperiled white abalone, after overcoming many hurdles, are now finally showing great potential for success.

If ever enough abalone do return to allow for fishing, it will need to be a very different kind of fishery—one that recognizes the biology and modern ecological context of the animals, and that also reconciles the deeper history of abalone use, encouraging the involvement of coastal tribes along with divers, marine conservation groups, and scientists to assure that broader perspectives are considered. To begin to rectify a long history of ignoring Native American interests in natural resources, the State of California has recently adopted policies that require the Fish and Game Commission and the Department of Fish and Wildlife to consult with tribes; this change promises to bring neglected indigenous viewpoints to bear in managing abalone in the future.[1]

With fisheries now closed, Californians are in a unique position to develop a different relationship with abalone based on stewardship. Already this is beginning to happen, with dedicated scientists and divers up and down the coast turning their passion for the ocean toward critical efforts to monitor marine environments, to develop local place-based knowledge and data, and to restore the kelp forests that abalone need.

Earlier generations took tremendous economic benefit and enjoyment from abalone. Now these modest animals need for us to give something back. No matter what has come before, the future of abalone depends on our adopting a conservation ethic that emphasizes not what we can get from these animals, but rather, what they need from us to thrive. The very future of the planet depends on us recognizing this same lesson writ large.

If the plight of an animal so beautiful, beloved, and closely intertwined with the lives and heritage of people along California's coast cannot compel us to pull together and act with courage and vision, we're in deep trouble. It's time to look into the brilliance of the abalone's shell and find inspiration to build a future that can accommodate our human needs for sustenance, beauty, and kinship for all earth's creatures.

Acknowledgments

The work of pulling together this book has been an incredible journey. Along the way I had the tremendous opportunity to meet and learn from many extraordinary people whose lives have intersected with abalone—so many of them generous with their knowledge, experience, and insights.

Thanks first to those who helped me learn about abalone evolution: Daniel Geiger at the Santa Barbara Museum of Natural History, Lindsey Groves at the Natural History Museum of Los Angeles County, and David Lindberg and the University of California at Berkeley. Though I was unable to include a chapter about evolution in the final manuscript owing to space constraints, I found a deeper understanding of abalone to be invaluable. Thanks also to the archaeologists who helped me learn what abalone shells can tell us about prehistory: Jon Erlandson, Gary Breschini, Chester King, Ray Corbett, and Lisa Thomas; their scholarship and conversation helped me make sense of the deep past. In addition, Arthur Vokes, Darrell Creel, Laura Nightengale, Tom Hester, John Greer, Matthew Taylor, Lisa Jackson, Ann Marie Early, Mary Beth Trubitt, Laura Kozuch, and Wes Tunnel helped me to verify the extent of the ancient shell trade.

I deeply appreciate Mati Waiya and Tima Lotah Link, who shared with me their personal perspectives on abalone loss to indigenous Californians and the importance of abalone to continuing Chumash culture. Thanks also to Rumsen basketmaker Linda Yamane for sharing her research and personal insights about the value and meaning of abalone to indigenous people of the Monterey area.

I could not have written this book without tapping the extensive knowledge and reflections of past, retired, and current biologists of the California Department of Fish and Wildlife, who have devoted their careers to the increasingly challenging project of conserving our marine fisheries, and abalone in particular: Earl Ebert, Richard Burge, Mel Odemar, Pete Haaker, Kristin Barsky, Konstantin Karpov, Ian Taniguchi, and Cynthia Catton. Thanks also to Laura Rogers-Bennett (also affiliated with UC Davis), who invited me to volunteer with creel surveys and to participate in a research day at Van Damme State Park so I could understand firsthand the state's ongoing abalone monitoring and research. In addition, Jim Moore, California's shellfish health expert, helped me understand the important details about withering syndrome.

Thanks also to the retired and current marine biologists at Channel Islands National Park who shared their experience and knowledge of that extraordinary place: Gary Davis, Dan Richards, David Kushner, and Stephen Whitaker. In particular, I am grateful to Dan Richards, who invited me to accompany him and his staff on an intertidal research trip so I could see the rare black abalone and better understand the animal's habitat and plight. He and his wife Sara also graciously welcomed me and my husband to join them on a personal abalone diving trip in northern California.

I appreciate NOAA scientist Melissa Neuman, who helped me understand the most recent chapters of the white abalone recovery effort, as did John Butler, John Hyde, and David Witting. In addition, Kristin Aquilino at the Bodega Marine Lab helped me understand the captive propagation effort. Heather Burdick of The Bay Foundation and Mike Schaadt at the Cabrillo Marine Aquarium shared their perspectives as white abalone recovery partners.

Thanks to Glenn VanBlaricom for allowing me to accompany him on a research trip to see recovering black abalone on San Nicolas and for sharing his depth of knowledge about not only *Haliotis* but also interactions with sea otters.

Thanks also to Paul Dayton, Ron Burton, Pete Raimondi, Kevin Lafferty, Lisa Crosson, John Pearse, Scoresby Shepherd, Marc Meyers, and Albert Yu-Min Lin for helping me understand the relevance of their scientific research to the abalone story. Retired Sea Grant specialist John Richards shared his perspective and information about sabellid worms, and Bryan Grummon helped me understand the optics of abalone iridescence.

Conservation-oriented sport divers Rocky Daniels, Steve Benavides, and Steve Campi shared inspiring stories about their campaign to end commercial abalone fishing. Thanks also to Brooke Halsey for telling me about his work as a district attorney and with wardens to avert the egregious problem of poaching. Representative Dan Hauser told me about his legislative effort to prevent commercialization of North Coast abalone.

Citizen-scientist and educator Nancy Caruso invited me to join her on a teaching day and also on a kelp restoration trip so I could learn about abalone in Orange County, for which I am grateful. And thanks to Josh Russo for telling me about his perspective as a sport diver and his efforts toward restoring northern California's kelp forests.

I appreciate that Steve Rebuck and Lad Handleman generously shared their knowledge, gave me insights on the perspective of commercial divers, and turned me on to Bob Kirby's memoir. Thanks also to Buzz Owens for sharing his unique perspective on abalone, from a career in commercial diving and mariculture but also from a personal passion for abalone hybrids.

Tim Thomas generously shared his research and thoughts about Pop Ernest and Monterey history and gave me the special opportunity to meet and learn from former Japanese American commercial diver Roy Hattori. Thanks also to Pat Sands for allowing me to include a photo of her grandfather Pop Ernest in this book.

In addition to people I interviewed directly, I have appreciated the many scholars and writers of books, reports, and articles whose works are cited in my endnotes, but especially those of Arthur McEvoy, Todd Braje, Les Field, and Scrap Lundy. Thanks also to Malcolm Margolin for his encouragement.

Many people told me their heartfelt, personal, soulful abalone stories, but I particularly appreciated the reflections of my friend and neighbor Gary Wickham.

Despite the knowledge, insights, and perspectives conveyed to me by all these people, all shortcomings, opinions, and errors that remain in this book are my own.

Thanks also to the endlessly helpful cadre of librarians and archivists: Carolyn Rainey and Peter Brueggeman at the Scripps Institution of Oceanography archives and library; Dennis Copeland at the Monterey Public Library's California History Room, Joan Parker at Moss Landing Marine Labs Library, Linda Johnson and Mary Quackenbush at the California State Archives and annex, Mike McCurdy and Angelica Illueca at the Witkin State Law Library, Joyce Winn at the National Sea Grant Library, and, at my local public library, Midge Hayes, who helped me borrow some obscure books through interlibrary loan. Pat Hathaway at the remarkable California Views Historical Photo Collection based in Monterey was also incredibly helpful.

Thanks to up-and-coming artist, illustrator, and writer Dominick Leskiw for his original abalone drawings and to my friend Teresa Bird for help with graphs used in the text.

Thanks to California friends who provided generous hospitality: Diana Selig and Meredith Rose, Erin and Blake Herron, Bill Avery, Ann Kelleher, Joaquin Feliciano, and Jennifer Price. Family members in California also provided hospitality and encouragement: Greg and Mary Bettencourt, Karen and Ted Lon, and especially, my aunt Birute Vileisis. Thanks also to my dear mother Janet Taylor for singing me a song each morning and to my sister Regina Krell for checking in on my progress.

Thanks to many other dear friends who have provided encouragement, inspiration, and ideas about writing, wildlife management, and marine science, especially Florence Williams, Vicki Graham, Vibeke Wilberg, Beth Jacobi, Linda Tarr, Mary Wahl, Tom Calvanese, and Larry Basch.

William Cronon gave me a lifetime's worth of inspiration and passion for writing environmental history, and my former editor and dear friend Jonathan Cobb gave me good advice when I needed it. I appreciate, too, the constructive comments provided by notable fisheries historian Carmel Finley and an anonymous reviewer.

Of course, this book would not have come to life without the support and enthusiasm of the Oregon State University Press staff, especially director Tom Booth; former acquisitions editor Mary Braun; current acquisitions editor Kim Hogeland; editorial, design, and production manager Micki Reaman; and marketing manager Marty Brown. Thanks also to copy editor Susan Campbell and proofreader Tracy Ann Robinson for their careful work.

Above all, thanks to my husband, author and photographer Tim Palmer, a wonderful partner in all of life's adventures. He has believed in me and the value of my work every single day of this journey. Without his abiding love and support, this book would have not been possible.

Notes on Sources

One of the most engaging parts of a historian's work is ferreting out the stories, details, and data in the written record that help us better understand the past. For this interdisciplinary history of abalone, references came from sources far and wide: from books, journals, and magazines in library stacks at UC Davis, Humboldt State University, Portland State University, and Scripps Institution of Oceanography to the modern bibliographic databases that make the scholarly literature of science and the humanities available to researchers; from traditional archives with boxes of yellowed papers to the remarkable digital libraries that increasingly make obscure historical sources more widely accessible.

For the first part of the book, I found these digital collections especially useful sources in archaeology, anthropology, ethnography, early European science, exploration, and colonization: UC Berkeley's George and Mary Foster Anthropological Library AnthroHub, University of California's eScholarship respository of open access publications, UC Press's e-books, Humboldt State University's Digital Library, the Smithsonian Libraries' Digital Collections, the Internet Archive, the eclectic Internet Sacred Text Archive, the HathiTrust Digital Library, Google Books, Gallica, ProQuest Dissertations and Theses, and the Wisconsin Historical Society's American Journeys digital library of eyewitness accounts.

For late nineteenth- and early twentieth-century California history, newspapers were invaluable references to understand local politics. Though I started out spending days reading microfiche at the California State Archives and Monterey Public Library, over the course of the nearly ten years it took me to research this book, many historical newspapers became available in digital form. For historical newspapers, I found the following online sources to be most useful: the California Digital Newspaper Collection, ProQuest Historical Newspapers, NewsBank, and the Monterey Public Library's online collection. The Monterey Public Library's California History Room was also a useful archive for materials related to that city's history.

Historical magazines were useful for understanding changing perceptions of abalone. Most late nineteenth- and early twentieth-century popular magazines must still be found in stacks or retrieved from the deep storage of libraries, but some are available online at the Internet Archive or Google

Books. Some historical books and journals that focus on natural history and science can be found online at the Biodiversity Heritage Library.

I found late nineteenth- and early to mid-twentieth-century federal and state government reports and historical publications, including those of the US Commission of Fish and Fisheries and the California Board of Fish Commissioners, at California State Library (CSL), and in UC Davis stacks, but also online at the Internet Archive, the Smithsonian Digital Library, Google Books, and the Biodiversity Heritage Library. In particular, the California Department of Fish and Game's *Fish Bulletins* can be found at California Digital Library's Calisphere, while the *California Fish and Game* journal can be found at the Internet Archive; both were useful for tracing growing scientific understanding of abalone and ongoing debates about fishery regulations. Useful for scientific context in the second half of the twentieth century, California Cooperative Ocean Fisheries Investigations reports can be found at its website.

State government reports, conference proceedings, and other materials related to the 1960s abalone–sea otter controversy can be found at the California State Library and also at the Moss Landing Marine Labs Library and Archive.

I found California Department of Fish and Game (CDFG) Cruise Reports and the *Kelp Forest Newsletter*, plus so-called gray literature and environmental documents from the 1990s, at the CDFG State Fisheries Library in Long Beach and at the Moss Landing Marine Labs Library; digital copies of some cruise reports have also been uploaded to the Aquatic Commons website. The National Sea Grant Library provided grant reports and publications funded by their agency, available either online or upon request.

I located early recreational diving magazines in the Scripps Institution of Oceanography Library vault and followed more recent history through dive club websites and online newsletters.

For research related to the California Fish and Game Commission (CFGC), the CDFG, and the legislature, I found meeting minutes, agency files, and bill files for much of the twentieth century at the California State Archives in Sacramento. These files also contained letters submitted by commercial and sport divers and organizations that provided useful insight into different perspectives on abalone issues. The Witkin State Law Library was also a helpful resource for specific details about legislation.

For CFGC meetings since 2003, video meeting minutes can be found at https://fgc.ca.gov/Meetings/Video. The most recent meetings, agendas, and summaries, including public comment letters, can also be found on the CFGC website. In addition, meeting agendas and summaries for the CFGC Marine Resources Committee can also be found on the CFGC website.

CDFW fisheries data, news updates, agendas and summaries of stake-holder meetings, and fishery management initiatives can be found at the CDFW marine region website, https://www.wildlife.ca.gov/Regions/Marine. In particular, the state's Abalone Recovery and Management Plan (ARMP), which includes a useful historical synopsis of California's abalone regulations, can be found at https://www.wildlife.ca.gov/Conservation/Marine/ARMP. Lectures presented at the CDFW's informative Conservation Lecture Series can be viewed online at https://www.wildlife.ca.gov/Conservation/Lectures.

Documents related to the management of federally endangered abalone species, such as listing notices, status reports, and recovery updates, can be found at the NOAA Fisheries website. Documents related to the threatened sea otter can be found at the US Fish and Wildlife Service and the US Geological Survey websites.

Beyond written documents, dozens of interviews were also important sources of information, perspective, and inspiration. Interviews are noted in the citations that follow, and the many scientists, divers, and others who were generous with their time and knowledge are also included in the acknowledgments.

Abbreviations

ARMP	Abalone Recovery and Management Plan, 2005
CDFG	California Division of Fish and Game (1928–1951)
CDFG	California Department of Fish and Game (1951–2012)
CDFW	California Department of Fish and Wildlife (2013–present)
CBFC	California State Board of Fish Commissioners (1870–1910)
CFGC	California Fish and Game Commission (1910–present)
CalCOFI	California Cooperative Ocean Fisheries Investigations
CSA	California State Archives
CSL	California State Library
CINP	Channel Islands National Park
NMFS	National Marine Fisheries Service
NOAA	National Oceanic and Atmospheric Administration
NPS	National Park Service
SCAN	Sonoma County Abalone Network
SIOA	Scripps Institution of Oceanography Library Archives
USDI	US Department of the Interior
USGS	US Geological Survey

Notes

INTRODUCTION

1 Daniel L. Geiger and Lindsey T. Groves, "Review of Fossil Abalone," *Journal of Paleontology* 73, no. 5 (1999): 872–885; Daniel Geiger, interview, April 16, 2010; Lindsey Groves, interview, June 29, 2010.

2 James A. Estes and Peter D. Steinberg, "Predation, Herbivory, and Kelp Evolution," *Paleobiology* 14, no. 1 (1988): 19–36; James A. Estes et al., "Evolution of Large Body Size in Abalones (*Haliotis*): Patterns and Implications," *Paleobiology* 31, no. 4 (2005): 591–606; David Lindberg, interview, April 9, 2010.

3 Geerat Vermeij, *Privileged Hands: A Remarkable Scientific Life* (New York: W. H. Freeman, 1997), 197–222.

4 Buzz Owen and Dwayne Dinucci, "A Brief History and Photo Study of the World's Six Largest *Haliotis* Shells," *Of Sea and Shore* 26, no. 4 (2004): 249–258.

CHAPTER 1

1 Jon M. Erlandson, "The Role of Shellfish in Prehistoric Economies, A Protein Perspective," *American Antiquity* 53, no. 1 (1988): 102–109.

2 Jon M. Erlandson et al., "Paleoindian Seafaring, Maritime Technologies, and Coastal Foraging on California's Channel Islands, *Science* 331 (2011): 1181–1185.

3 Jon Erlandson, interview, October 29, 2013; Todd J. Braje et al., "Human Impacts on Nearshore Shellfish Taxa: A 7,000-Year Record from Santa Rosa Island, California," *American Antiquity* 72 (2007): 741.

4 Braje et al., "Human Impacts," 740–741.

5 Jon M. Erlandson et al., "Human Impacts on Ancient Shellfish: A 10,000-Year Record from San Miguel Island, California," *Journal of Archaeological Science* 35 (2008): 2145; Braje et al., "Human Impacts," 747–748; Clement W. Meighan, "The Little Harbor Site, Catalina Island: An Example of Ecological Interpretation in Archaeology," *American Antiquity* 24, no. 4 (1959): 402.

6 Michael A. Glassow et al., "Red Abalone Collecting and Marine Water Temperature during the Middle Holocene Occupation of Santa Cruz Island, California," *Journal of Archaeological Science* 39 (2012): 2574; Torben C. Rick et al., "Early Red Abalone Shell Middens, Human Subsistence, and Environmental Change on California's Northern Channel Islands," *Journal of Ethnobiology* 39, no. 2 (2019): 205.

7 Braje et al., "Human Impacts," 735–756; John A. Robbins et al., "A 7000-Year Sea Surface Temperature Record from CA-SRI-147, Santa Rosa Island, California, USA," *The Holocene* 23, no. 7 (2013): 1008–1016; John Sharp, "Shellfish Analysis from a Santa Cruz Island Red Abalone Midden: Re-evaluating the Marine Cooling Hypothesis," *Fifth California Islands Symposium* (Santa Barbara Museum of Natural History, 2002), 563–564; Roy A. Salls, "Early Holocene Maritime Adaptation at Eel Point, San Clemente Island," in *Hunter-Gatherers of Early Holocene Coastal California*, Perspectives in California Archaeology, vol. 1, ed. Jon Erlandson and Roger Colten (UCLA Institute of Archaeology, 1991), 60–77.

8 Jon M. Erlandson et al., "Sea Otters, Shellfish, and Humans: 10,000 Years of Ecological Interactions on San Miguel Island, California," *Sixth California Islands Symposium* (Institute for Wildlife Studies, 2005), 58–69; William J. Douros, "Prehistoric Predation on Black Abalone by Chumash Indians and Sea Otters," *Third California Islands Symposium* (NPS, 1993), 560–564.

9 Torben C. Rick and Jon M. Erlandson, "Coastal Exploitation," *Science* 325 (2009): 953; Dana Lepofsky et al., "Ancient Shellfish Mariculture on the Northwest Coast of North America," *American Antiquity* 80, no. 2 (2015): 237–259.

10 Jon M. Erlandson et al., "Sea Otters, Shellfish, and Humans," 58–69; Terry L. Jones et al., "Towards a Prehistory of the Southern Sea Otter (*Enhydra lutris nereis*)," in *Human Impacts on Seals, Sea Lions, and Sea Otters: Integrating Archaeology and Ecology in the Northeast Pacific*, ed. Todd J. Braje and Torben C. Rick (University of California Press, 2011), 266–267.

11 Torben C. Rick et al., "From Pleistocene Mariners to Complex Hunter-Gatherers: The Archaeology of the California Channel Islands," *Journal of World Prehistory* 19 (2005): 187, 194, 217; Jon M. Erlandson et al., "Beads, Bifaces, and Boats: An Early Maritime Adaptation on the South Coast of San Miguel Island, California," *American Anthropologist* 107, no. 4 (2005): 679–680; Jon Erlandson, correspondence, September 6, 2019.

12 Adrian R. Whitaker and Brian F. Byrd, "Boat-Based Foraging and Discontinuous Prehistoric Red Abalone Exploitation along the California Coast," *Journal of Anthropological Archaeology* 31 (2012): 196–214; Erlandson et al., "Human Impacts on Ancient Shellfish," 2147; Todd J. Braje et al., "Fishing from Past to Present: Continuity and Resilience of Red Abalone Fisheries on the Channel Islands, California," *Ecological Applications* 19, no. 4 (2009): 910.

13 Chester D. King, "The Evolution of Chumash Society: A Comparative Study of Artifacts Used in Social System Maintenance in the Santa Barbara Channel Region before A.D. 1804," PhD diss., UC Davis, 1981.

14 Georgia Lee, *The Portable Cosmos: Effigies, Ornaments, and Incised Stone from the Chumash Area* (Ballena Press, 1981), 23; Travis Hudson and Thomas Blackburn, *The Material Culture of the Chumash Interaction Sphere* (Ballena Press, 1982), 48, 103.

15 Lee, *Portable Cosmos*, 23; Roberta S. Greenwood, "9,000 Years of Prehistory at Diablo Canyon," *San Luis Obispo County Archaeological Society Occasional Papers* (1972): 35–36; Travis Hudson et al., eds., *The Eye of the Flute: Chumash Traditional History and Ritual, as Told by Fernando Librado Kitsepawit to John P. Harrington* (Santa Barbara Museum of Natural History, 1977), 42.

16 E. W. Gifford, "Californian Shell Artifacts," *Anthropological Records* 9, no. 1 (1947): 5–6, 50–61; James T. Davis, "Trade Routes and Economic Exchange among the Indians of California," *University of California Archaeological Survey* 54 (1961): 11, passim.

17 Robert E. C. Stearns, "Ethno-conchology—A Study of Primitive Money," *Report of the U.S. National Museum*, vol. 2 (GPO, 1887), 320–326; Robert F. Heizer, ed. "The Northern California Indians: A Reprinting of 19 Articles by Stephen Powers on California Indians Originally Published 1872–1877," *Contributions of the University of California Archaeological Research Facility* 25 (1975); Jerry Marten, *Shell Game: A True Account of Beads and Money in North America* (Mercury House, 1996).

18 Ray Corbett, interview, March 27, 2014.

19 Lynn Gamble, *The Chumash World at European Contact: Power, Trade, and Feasting among Hunter Gatherers* (University of California Press, 2008), 183; Jon M. Erlandson et al., *A Canyon through Time: Archaeology, History and Ecology of the Tecolote Canyon Area, Santa Barbara County, California* (University of Utah Press, 2008), 71–72.

20 Gamble, *Chumash World*, 180, 185; Hudson and Blackburn, *Material Culture*, 103.

21 Steven J. Schwartz, "Some Observations on the Material Culture of the Nicoleño," *Sixth California Island Symposium* (Institute for Wildlife Studies, 2005), 83–91; Lisa Thomas, curator, San Nicolas Island, interview, October 24, 2015.

22 King, "Evolution of Chumash Society," 157.

23 King, "Evolution of Chumash Society," 158–159; Lynn H. Gamble et al., "An Integrative Approach to Mortuary Analysis: Social and Symbolic Dimensions of Chumash Burial Practices," *American Antiquity* 66, no. 2 (2001): 187, 200, 208; Phil C. Orr, "Exceptional Burial in California," *Science* 94 (December 1941): 539; Chester King, interview, April 2, 2014.

24 Barbara Lee Jones, "Mythic Implications of Faunal Assemblages from Three Ohlone Sites," master's thesis, San Francisco State University, 2010.

25 Lindee R. Grabouski, "Smoke and Mirrors: A History of NAGPRA and the evolving U.S. View of the American Indian," master's thesis, University of Nebraska, 2011; Kate Wong,

"Archaeologists and Native Americans Team Up to Interpret the Past, Shape the Future," *Scientific American* blog, March 31, 2011.

26 Phil C. Orr, *Prehistory of Santa Rosa Island* (Santa Barbara Museum of Natural History, 1965), 125, 134.

27 Jones, "Mythic Implications," 47.

28 Orr, *Prehistory*, 138–139, 222, 228; Gamble et al., "Integrative Approach," 201; Terisa Green, "And Both Parents Are Fond of Their Children: Reinterpretation of Wealthy Chumash Child Burials," *Backdirt, Annual Review of the Cotsen Institute of Archaeology* (Spring/Summer 1999).

29 Thomas C. Blackburn, ed. *December's Child: A Book of Chumash Oral Narratives, Collected by J.P. Harrington* (University of California Press, 1975), 100; Brian D. Haley and Larry R. Wilcoxon, "Point Conception and the Chumash Land of the Dead: Revisions from Harrington's Notes," *Journal of California and Great Basin Anthropology* 21, no. 2 (1999): 213–235.

30 Frank Latta, *Handbook of the Yokuts Indians*, 2nd ed. (Bear State Books, 1977), 321–322. In their Image Ceremonies, Kumeyaay ancestors in the present-day San Diego area also used abalone shell eyes in the masks they made to mourn and honor the dead; see Constance Goddard Dubois, "Religious Ceremonies and Myths of the Mission Indians," *American Anthropologist* 7, no. 4 (1905): 625–626.

31 Pliny Earle Goddard, "Chilula Texts," *University of California Publications in American Archaeology and Ethnology* 10, no. 7 (1914): 289–379; Richard Keeling, *Cry for Luck: Sacred Song and Speech among the Yurok, Hupa, and Karok Indians of Northwestern California* (University of California Press, 1992), 277.

32 Orr, *Prehistory*, 189.

33 Robert F. Heizer, "California Indian Linguistic Records: The Mission Indian Vocabularies of H. W. Henshaw," *Anthropological Records* 15, no. 2 (University of California Press, 1955), 152, 155; Hudson and Blackburn, *Material Culture*, 110. A Kumeyaay descendant reported a similar use for abalone scratchers during the seclusion period before a girl's coming of age ceremony; see Constance Goddard DuBois, "The Religion of the Luiseño Indians of Southern California," *University of California Publications in American Archaeology and Ethnology* 8, no. 3 (1908): 93.

34 Les W. Field, *Abalone Tales: Collaborative Explorations of Sovereignty and Identity in Native California* (Duke University Press, 2008), 117–119; Pliny Earle Goddard, "Life and Culture of the Hupa," *University of California Publications in American Archaeology and Ethnography* 1 (1903): 53–54.

35 Hudson and Blackburn, *Material Culture*, 26; Jones, "Mythic Implications," 31.

36 Amy J. Gilreath, "Rock Art in the Golden State," in *California Prehistory: Colonization, Culture, and Complexity*, ed. Terry L. Jones and Kathryn A. Klar (AltaMira Press, 2007), 277.

37 Anne H. Gayton, "Yokuts and Western Mono Ethnography," *Anthropological Records* 10, no. 1 (1948): 37–38, cited in Lee, *Portable Cosmos*, 23.

38 Don Laylander, "Early Ethnographic Notes from Constance Goddard Dubois on the Indians of San Diego County," *Journal of California and Great Basin Anthropology* 26, no. 2 (2006): 205–214.

39 James A. Bennyhoff and Richard A. Hughes, "Shell Bead and Ornament Exchange Networks between California and the Western Great Basin," *Anthropological Papers of the American Museum of Natural History* 64, no. 2 (1987): 82; Arthur W. Vokes and David A Gregory, "Exchange Networks for Exotic Goods in the Southwest and Zuni's Place in Them," in *Zuni Origins: Toward a New Synthesis of Southwestern Archaeology*, ed. David Gregory and David Wilcox (University of Arizona Press, 2007), 334–337; Arthur Vokes, interview, March 27, 2013.

40 Michael D. Cook, ed., "The Cultural Resources of Quail Creek," report, vol. 1 (Westland Resources, 2006), fig. 4.2 (author collection); William Hoyt Smith, "Trade in Molluskan Religiofauna between the Southwestern United States and Southern California," PhD diss., University of Oregon, 2002, 207–209; Wesley Jernigan, "The Cloud and the Shell,"

Arizona Highways 57, no. 10 (1981): 41; Neil M. Judd, "The Material Culture of Pueblo Bonito," *Smithsonian Miscellaneous Collections* 124 (1954): 88.

41 Arthur W. Vokes, "The Shell Ornament Assemblage," in *Tonto Creek Archaeological Project: Life and Death along Tonto Creek,* Anthropological Papers 24 (Center for Desert Archaeology, 2001): 353–420; Wesley Jernigan, *Jewelry of the Prehistoric Southwest* (School of American Research, 1978), 103; Vokes, interview, 2013.

42 Lisa W. Huckell, "The Shell Assemblage from Coffee Camp," in *Archaic Occupation on the Santa Cruz Flats: The Tator Hills Archaeological Project,* ed. Carl D. Halbirt and T. Kathleen Henderson (Northland Research, 1993), 305–316; Vokes, interview, 2013.

43 Vokes and Gregory, "Exchange Networks," 336–337; Judd, "Material Culture," 94–95.

44 Pliny Earle Goddard, "Navajo Texts," *Anthropological Papers of the American Museum of Natural History* 34 (1933): 147.

45 Elsie Clews Parsons, "War God Shrines of Laguna and Zuñi," *American Anthropologist,* new series 20, no. 4 (1918): 397, n2; Ruth L. Bunzel, "Zuñi Katcinas: An Analytical Report," *Forty-Seventh Annual Report of the Bureau of American Ethnology* 1929–1930 (GPO, 1932), 871, 1032.

46 Darrell G. Creel, "Ceremonial Cave: An Overview of Investigations and Contents," in *The Hueco Mountain Cave and Rock Shelter Survey: A Phase 1 Baseline Inventory in Maneuver Area 2D on Fort Bliss, Texas,* ed. Federico A. Almarez and Jeff D. Leach, Archaeological Technical Reports no. 10; C. B. Cosgrove, "Caves of the Upper Gila and Hueco Areas in New Mexico and Texas," *Papers of the Peabody Museum of American Archaeology and Ethnology, Harvard University* 24, no. 2 (Peabody Museum, 1947).

47 Reuben G. Thwaites, ed., *Original Journals of the Lewis and Clark Expedition, 1804–1806,* vol. 2 (Dodd, Mead, 1904), 378; see also vol. 3, pp. 328, 345, 352, and vol. 4, pp. 187, 191.

48 Though previous accounts have suggested that abalone was traded as far east as the Mississippi River, I could find no specimens verified by modern archaeologists. The provenance of one black abalone shell reported at Spiro Mound in eastern Oklahoma by Laura Kozuch, "Marine Shells from Mississippian Archaeological Sites" (PhD diss., University of Florida Gainesville, 1998) has since been questioned by the author, who now believes that that shell may have been misfiled; Laura Kozuch correspondence, February 18, 2013.

49 Torben C. Rick et al., "On the Antiquity of the Single-Piece Shell Fishhook," *Journal of Archaeological Science* 29 (2002): 933–942; Jon M. Erlandson et al., "Fishing up the Food Web? 12,000 Years of Maritime Subsistence and Adaptive Adjustments on California's Channel Islands," *Pacific Science* 63, no. 4 (2009): 711–724.

50 Jeanne Arnold and Anthony P. Graesch, "The Evolution of Specialized Shellworking among the Island Chumash," in *The Origins of a Pacific Coast Chiefdom: The Chumash of the Channel Islands* (University of Utah Press, 2001), 108–109.

51 Rene L. Vellanoweth et al., "Middle and Late Holocene Maritime Adaptations on Northeastern San Miguel Island, California," *Fifth California Islands Symposium* (Santa Barbara Museum of Natural History, 2002), 612; Torben C. Rick et al., "From Pleistocene Mariners to Complex Hunter-Gatherers: The Archaeology of the California Channel Islands," *Journal of World Prehistory* 19 (2005): 212–215; Roy A. Salls, "The Prehistoric Fishery of San Clemente Island," *Pacific Coast Archaeological Society Quarterly* 36, nos. 1/2 (2000): 60.

52 Rick et al., "From Pleistocene Mariners," 209; Arthur Woodward, "Shell Fish Hooks of the Chumash," *Bulletin of the Southern California Academy of Sciences* 28, no. 3 (1930): 41–42; J. P. Harrington, "Exploration of the Burton Mound at Santa Barbara, California," *Forty-Fourth Annual Report of the Bureau of American Ethnology, 1926–27* (GPO, 1928), 139–146; Hudson and Blackburn, *Material Culture,* 172.

53 "T'aya" (Chumash) in Haley and Wilcoxon, "Point Conception," 223; Mati Waiya, interview, April 7, 2011; "A'ulun" (Costanoan/Ohlone) in J. P. Harrington, "Indian Words in Southwest Spanish, Exclusive of Proper Nouns," *Plateau* 17 (1944): 28, 37; Linda Yamane, interview, March 26, 2011; "Ah'-wook" (Miwok) in C. Hart Merriam, ed., *The Dawn of the World: Myths and Weird Tales Told by the Mewan (Miwok) Indians of California* (Arthur H. Clarke, 1910), 201; "Wil" (Kashaya Pomo) in Cora Alice DuBois, "1870 Ghost Dance," *University of California Anthropological Records* 3, no. 1 (1939): 97; Field,

Abalone Tales, 66; *"Hiwo't"* (Wiyot) from Myth (Latsik) of Abalone (Hiwo't), #24-2540, in The Alfred L. Kroeber Collection of American Indian Sound Recordings, Phoebe A. Hearst Museum of Anthropology, UC Berkeley, http://cla.berkeley.edu/item/13228; *"Yer'erner"'* (Yurok) from Yurok language project and "The shell's boat dance into the ocean" in Robert Spott and A. L. Kroeber, "Yurok Narratives," *University of California Publications in American Archaeology and Ethnology* 35 (1942): 249–250; *"Yuxthâran"* (Karuk) Karuk Dictionary by William Bright and Susan Gehr, lexicon ID #7263, http://linguistics.berkeley.edu/~karuk/karuk-dictionary.php; *"Xosa:k"'* (Hupa) in Na:tinixwe Mixine:whe', *Hupa Language Dictionary*, 2nd ed. (Hoopa Valley Tribal Council, 1995), 1; *"Ila'k'waasht'i"* (Tolowa) from Me'laashne Loren Bommelyn, *Now You're Speaking Tolowa* (Center for Indian Community Development, Humboldt State University, 1995), 3.

54 Merriam, *Dawn of the World*, 201.

55 Alfred L. Kroeber, "Wishosk Myths," *Journal of American Folklore* 18, no. 69 (1905): 90–91, 93–94. This story was told to Kroeber by a Wiyot man identified only as "Bob," but it's worth noting that Kroeber disparaged Bob's storytelling as "incoherent statements." Few people spoke the Wiyot language at the time, so Kroeber was anxious to record what he could, yet one has to wonder about what was lost in his translation.

56 Florence Silva, interview by Les Field, in *Abalone Tales*, 66, 76.

57 Les Field, "Abalone Tales: California Indian Natural Resource Management," *News from Native California* (Winter 2002/2003): 32–33.

58 Julian Lang, "Reflections on the Iridescent One," in Field, *Abalone Tales*, 84–106; Callie Lara, interview by Les Field, in *Abalone Tales*, 112–114.

59 "Abalone: A Wiyot Tale by Della Henry Prince," *News from Native California* (Winter 2002/2003), 22, reprinted from Karl V. Teeter and John D. Nichols, *Wiyot Handbook II: Interlinear Translation and English Index*, Algonquin and Iroquoian Linguistics Memoirs 11 (University of Manitoba, Department of Linguistics, 1993), 1–4.

60 Lang, "Reflections on the Iridescent One," 96.

61 Lyn Risling, "Abalone Woman and Dentalium Man," *News from Native California* (Summer 2014): 22.

62 Bradley Marshall, interview by Les Field, in *Abalone Tales*, 76, 121–129.

63 Intertribal Gathering and Elder's Dinner, November 11, 2011; http://www.nativewomenscollective.org/regaliastoriesabalone.html.

CHAPTER 2

1 *First from the Gulf to the Pacific: The Diary of the Kino-Atondo Peninsular Expedition (Dec. 14, 1684–Jan. 13, 1685)*, transcribed, translated, and edited by W. Michael Mathes (Dawson's Bookshop, 1969), 45.

2 Ronald L. Ives, "The Quest of the Blue Shells," *Arizoniana* 2, no. 1 (1961): 3–7; Herbert E. Bolton, *Kino's Historical Memoir of Pimería Alta, a contemporary account of the beginnings of California, Sonora, and Arizona, by Father Eusebio Francisco Kino, S.J. pioneer, missionary, explorer, cartographer and ranchman, 1683–1711*, vol. 1 (Arthur Clark, 1919), 195–196, 230.

3 Bolton, *Kino's Historical Memoir*, 330–333.

4 Salmerón Zárate, "Journey of Oñate to California by Land" [1626], cited in Herbert E. Bolton, ed., *Spanish Exploration in the Southwest, 1542–1706*, vol. 17 (Charles Scribner's, 1916), 273–276.

5 Jack Hicks, *The Literature of California: Native American Beginnings to 1945* (University of California Press, 2000), 76.

6 Kevin Starr, *California: A History* (Random House, 2007), 30–31; Natale A. Zappia, *Traders and Raiders: The Indigenous World of the Colorado Basin, 1540–1859* (University of North Carolina Press, 2014), 53–57.

7 Bolton, *Kino's Historical Memoir*, 237.

8 Bolton, *Kino's Historical Memoir*, 237.

9 Bolton, *Kino's Historical Memoir*, 231, 289, 310, 317, 323, 342, 352.

10 Bolton, *Kino's Historical Memoir*, 242.

11 Bolton, *Kino's Historical Memoir*, 251.

12 Bolton, *Kino's Historical Memoir*, 249–250; Ronald L. Ives, "Kino's Exploration of the Pinacate Region," *Journal of Arizona History* 7, no. 2 (1966): 71–73.

13 Bolton, *Kino's Historical Memoir*, 343.

14 Bolton, *Kino's Historical Memoir*, 284, 287, 307–313, 318–319; Harlan Hague, "The Search for a Southern Overland Route to California," *California Historical Society* 55, no. 2 (1976): 153.

15 Hague, "Search," 155.

16 Steven Hackel, *Children of Coyote, Missionaries of Saint Francis: Indian-Spanish Relations in Colonial California, 1769–1850* (University of North Carolina Press, 2005); Benjamin Madley, *An American Genocide: The United States and the California Indian Catastrophe* (Yale University Press, 2016), 16–41.

17 Sherburne F. Cook, *The Population of the California Indians, 1769–1970* (University of California Press, 1976), 20–43; and "Historical Demography," in *Handbook of North American Indians*, Robert F. Heizer, ed. (Smithsonian Institution, 1978), 91–93.

18 Robert Howard Jackson and Edward Castillo, *Indians, Franciscans, and Spanish Colonization: The Impact of the Mission System on California Indians* (University of New Mexico Press, 1996), 32–35; Hackel, *Children of Coyote*, 84, 263; "Chumash Tabernacle," in *Galleons and Globalization: California Mission Arts and the Pacific Rim,* exhibit at University of San Francisco Library, August–December 2010; Kent G. Lightfoot, *Indians, Missionaries, and Merchants: The Legacy of Colonial Encounters on the California Frontiers* (University of California Press, 2005), 97, 109; Russell K. Skowronek, "Sifting the Evidence: Perceptions of Life at the Ohlone (Costanoan) Missions of Alta California," *Ethnohistory* 45, no. 4 (1998): 675–798.

19 H. R. Wagner, "The Voyage of Pedro de Unamuno to California in 1587," *California Historical Society Quarterly* 2, no. 2 (1923): 150, 157; Bolton, *Spanish Exploration*, 119–120.

20 "Journal of Fray Juan Crespi kept during the voyage of the Santiago, account dated October 5, 1774," in *Documents from the Sutro Collection, Publications of the Historical Society of Southern California* 2, no. 1 (1891): 132, ed. George Butler Griffen.

21 Adele Ogden, "The Californias in Spain's Pacific Otter Trade, 1775–1795," *Pacific Historical Review* 1, no. 4 (1932): 444. According to Ogden, Spanish sailors started to barter for West Coast sea otter pelts before the Russians and English. At the time, the Russian fur trade centered on the Aleutian Islands.

22 Ogden, "The Californias," 445.

23 Jose Mariano Mozino, *Noticias de Nutka: an account of Nootka Sound in 1792*, trans. and ed. Iris Wilson Engstrand, 2nd ed. (University of Washington Press, 1991), 19; Franz Boas, "Ethnology of the Kwakiutl," *Thirty-Fifth Annual Report of the Bureau of American Ethnology, 1913–1914*, part 2 (GPO, 1915), 1069, 1271, 1273, 1274; Franz Boas, "Tshimsian Myths," *Thirty-First Annual Report of the Bureau of American Ethnology, 1909–1910* (GPO, 1911), 53, 188, 231–232, 263, 378.

24 Adele Ogden, *The California Sea Otter Trade, 1784–1848* (University of California Press, 1941), 29.

25 Ogden, "The Californias," 445–446; Robert F. Heizer, "Indians and Abalone Shells," *California Monthly* 75, no. 9 (1965): 14–15.

26 Ogden, "The Californias," 445–468.

27 James R. Gibson, *Otter Skins, Boston Ships, and China Goods: The Maritime Fur Trade of the Northwest Coast, 1785–1841* (University of Washington Press, 1992), 273–278.

28 Ogden, *California Sea Otter Trade*, 140–182.

29 Douglas P. DeMaster et al., "Estimating the Historical Abundance of Sea Otters in California," *Endangered Species Update* 13, no. 12 (1996): 79–81; Kristin L. Laidre et al., "An Estimation of Carrying Capacity for Sea Otters along the California Coast," *Marine Mammal Science* 17, no. 2 (2001): 294. These sources base their estimates on habitat carrying capacity. The 1996 study estimated a historic population of 13,500 sea otters in California; the 2001 study refined the estimate to 15,941.

30 Linda Yamane (Rumsen descendent) interview, March 26, 2011; Alexander Taylor, "California Notes—The Aulone Shell-Fish," *Sacramento Daily Union*, November 21, 1856,

CSA, newspaper clipping file; J. P. Harrington, "Indian Words in Southwest Spanish, Exclusive of Proper Nouns," *Plateau* 17 (1944): 28, 37. Ethnographer Alfred Kroeber errantly suggested that the Ohlone word *a'ulun* derived from the Spanish, but historic records indicate that it was the other way around. The Spanish had used other words to describe abalone shells, such as *conchas de orejas* (ear shells) and *conchas de Monterey*; A. L. Kroeber, "The Chumash and Costanoan Languages," *University of California Publications in American Archaeology and Ethnography* 9 (1910): 257; also, Donald T. Clark, *Monterey County Place Names: A Geographical Dictionary* (1991), 406, says that Point Aulones can be found on maps dating back to 1856.

31 Taylor, "California Notes," 4.

32 Richard Henry Dana, *Two Years before the Mast* (D. Appleton, 1899), 107. When Henry Richard Dana arrived in Monterey in 1834, he learned that "the place was celebrated for shells."

CHAPTER 3

1 M. R. Snow and A. Pring, "The Mineralogical Microstructure of Shells, Part 2: The Iridescence Colors of Abalone Shells," *American Mineralogist* 90 (2005): 1706.

2 Margaret Maria Gordon, *The Home Life of Sir David Brewster* (Edmonston and Douglas, 1870), 97, 154, 259.

3 David Brewster, "On New Properties of Light Exhibited in the Optical Phenomena of Mother of Pearl, and Other Bodies to Which the Superficial Structure of That Substance Can Be Communicated," *Philosophical Transactions of the Royal Society of London* 104 (1814): 398.

4 Brewster, "On New Properties," 406.

5 Brewster, "On New Properties," 406.

6 Brewster, "On New Properties," 409.

7 Brewster, "On New Properties," 410.

8 Brewster, "On New Properties," 415.

9 "Causes of Color," web exhibit, http://www.webexhibits.org/causesofcolor/7A.html, accessed July 2, 2010.

10 "Causes of Color."

11 Snow and Pring, "Mineralogical Microstructure," 1706.

12 "Causes of Color."

13 Snow and Pring, "Mineralogical Microstructure," 1707.

14 Andrew R. Parker, "The Diversity and Implications of Animal Structural Colours," *Journal of Experimental Biology* 201 (1998): 2343–2347; Shuichi Kinoshita and Shinya Yoshioka, "Structural Colors in Nature: The Role of Regularity and Irregularity in the Structure," *ChemPhysChem* 6 (2005): 1442–1459.

15 Stephen J. Gould and Richard C. Lewontin, "The Spandrels of San Marco and the Panglossian Paradigm: A Critique of the Adaptationist Programme," *Proceedings of the Royal Society of London* B 205, no. 1161 (1979): 581–598.

16 Geerat J. Vermeij, *Evolution and Escalation: An Ecological History of Life* (Princeton University Press, 1987), 191–237; G. J. Vermeij, *Privileged Hands: A Remarkable Scientific Life* (Freeman, 1997), 197–222.

17 Marc Meyers, interview, April 27, 2010; Albert Yu-Min Lin, interview, April 27, 2010.

18 Philip Ball, "Life's Lessons in Design," *Nature* 409 (January 2001): 414; Peter Westbroek and Frederic Marin, "A Marriage of Bone and Nacre," *Nature* 392 (April 1998): 861; Helena Oh, "A Sea Change for Artificial Bones?" *BusinessWeek*, February 20, 2006; Matthew A. Carson and Susan A. Clarke, "Bioactive Compounds from Marine Organisms: Potential for Bone Growth and Healing," *Marine Drugs* 16 (2018): 340.

CHAPTER 4

1 Sandy Lydon, *Chinese Gold: The Chinese in the Monterey Bay Region* (Capitola, 1985), 20–24, 31–35.

2 "Abalone Meat—A New Article of Food," *San Diego Union*, October 5, 1871; Todd J. Braje and Jon M. Erlandson, "Historical Abalone Fishers on San Miguel Island, California,"

Proceedings of the Society for California Archaeology 19 (2006): 23; A. L. "Scrap" Lundy, *The California Abalone Industry: A Pictorial History* (Best, 1997), 5.

3   *The Finishing Touch, Lacquer,* exhibit text, E & J Frankel, New York, November 16–December 21, 1985; James C. Y. Watt and Barbara Brennan Ford, *East Asian Lacquer: The Florence and Herbert Irving Collection* (Metropolitan Museum of Art, 1991), 124–138.

4   Zhenping Wang, *Ambassadors from the Islands of Immortals: China-Japan Relations in the Han-Tang Period* (University of Hawai'i Press, 2005), 1–2, 7–9; Kenneth Ruddle, *Administration and Conflict Management in Japanese Coastal Fisheries,* FAO Fisheries Technical Paper 273 (1987), ch. 1, http://www.fao.org/3/T0510E/T0510E00.htm.

5   Deh-Ta Hsiung and Nina Simonds, *The Food of China* (Allen and Unwin, 2001), 9, 72; Dorothy Perkins, *Encyclopedia of China* (Routledge, 2013), 353.

6   Jacqueline M. Newman, *Food Culture in China* (Greenwood Press, 1994), 192.

7   John Branch, "Prized but Perilous Catch," *New York Times,* July 25, 2014.

8   Jacqueline M. Newman, "Abalone," *Flavor and Fortune* 9, no. 3 (2002): 5, 24; Dan Bensky and Andrew Gamble, *Chinese Herbal: Medicine Materia Medica,* rev. ed. (Eastbound Press, 1993), 425–426.

9   Lundy, *California Abalone Industry,* 5; P. G. Albrecht, "A Comparative Study of Certain Marine Mollusks of the Pacific Coast" (PhD diss., Stanford University, 1917), cited in Keith W. Cox, "California Abalones, Family Haliotidae," *Fish Bulletin* 118 (1962): 76.

10 "Chinese Shell Fishery at Monterey," *Daily Alta California,* May 20, 1853.

11 Lydon, *Chinese Gold,* 31; "Chinese Shell Fishery at Monterey," *Daily Alta California,* May 20, 1853.

12 Lundy, *California Abalone Industry,* 5.

13 Arthur F. McEvoy, "In Places Men Reject: Chinese Fishermen at San Diego, 1870–1893," *Journal of San Diego History* 23, no. 4 (1977): 3; "The Aulone," *Monterey Weekly Herald,* October 3, 1874; "Commercial and Financial," *Daily Alta California,* October 13, 1860; "Queer Articles of Diet," *Los Angeles Times,* May 7, 1887.

14 "A Valuable Mollusk: Abalone Shells and Shell Meat—China's Dainty Dish," *San Francisco Bulletin,* July 9, 1885.

15 "Rambles in California," *New York Times,* April 13, 1874; "Shellfish for San Franciscans," *Daily Alta California,* July 28, 1879.

16 "Commercial and Financial," *Daily Alta California,* October 13, 1860.

17 "The Chinese," *New York Times,* January 3, 1870; McEvoy, "In Places," 5.

18 "Commercial and Financial," *Daily Alta California,* October 13, 1860, 4; Lydon, *Chinese Gold,* 32; Lundy, *California Abalone Industry,* 7–8; Ernest Ingersoll, "The Abalone Fishery," in *The Fisheries and Fishery Industry of the United States,* ed. George Brown Goode, Sec. 5, Vol. 2, US Commission of Fish and Fisheries (GPO, 1887), 623; Linda Bentz and Robert Schwemmer, "The Rise and Fall of Chinese Fisheries in California," in *The Chinese in America: A History from Gold Mountain to the New Millennium,* ed. Susie Lan Cassel (Rowman/AltaMira, 2002), 149.

19 Lundy, *California Abalone Industry,* 9.

20 Robert E. C. Stearns, "Ethno-conchology—A Study of Primitive Money," *Report of the U.S. National Museum,* vol. 2 (GPO, 1887), 256.

21 Mrs. M. Burton Williamson, "The Haliotis or Abalone Industry of the Californian Coast: Preservative Laws," *Annual Publication of the Historical Society of Southern California* 7 (1906): 30; Charles H. Stevenson, "Economic Uses for Shells," *Scientific American,* supplement 54 (July 1902): 22201.

22 *Daily Alta California,* supplemental sheet, January 9, 1861; Ingersoll, "Abalone Fishery," 624. To estimate the extent of abalone take during the first rush, I searched export reports from San Francisco as reported in early newspapers through the California Digital Newspaper Collection. Because records were inconsistent, and the weights of sacks, bags, and shells are not well defined, it's not possible to derive a wholly reliable estimate. That said, using contemporary references, I was able to roughly calculate more than ten thousand sacks—more than one million pounds of dried abalone—which means a floor estimate of up to five million intertidal and nearshore abalone removed between 1860 and

1870. Using Ingersoll's factor for converting dry to wet weights, and then CDFG's weights of animals, more than two million animals were taken in 1879 alone.

23 Lydon, *Chinese Gold*, 33.

24 Arthur F. McEvoy, *The Fisherman's Problem: Ecology and Law in California Fisheries, 1850–1980* (Cambridge University Press, 1986), 104.

25 Ann Vileisis, *Kitchen Literacy: How We Lost Knowledge of Where Food Comes From and Why We Need to Get It Back* (Island Press, 2007), 151–155; "The End of the Buffalo: How the Mighty Herds Have Been Exterminated," *New York Times*, December 26, 1884.

26 "The Chinese Fishermen of the Pacific Coast, from Notes by David Starr Jordan," in *The Fish and Fishery Industries of the United States*, ed. George Brown Goode, Section 4, US Commission of Fish and Fisheries (GPO, 1887), 40; David Starr Jordan, *The Days of a Man: Being Memories of a Naturalist, Teacher, and Minor Prophet of Democracy* (World Book, 1922), 204; David Starr Jordan, "Fisheries of California," *Overland Monthly* 20 (November 1892): 473; J. W. Collins, "Report on the Fisheries of the Pacific Coast of the United States," in *US Commission of Fish and Fisheries, Report of the Commissioner for 1888*, part 16 (GPO, 1892), 26; McEvoy, *Fisherman's Problem*, 96–97.

27 Jordan, "Fisheries," 473; McEvoy, *Fisherman's Problem*, 6.

28 Jordan, "Chinese Fishermen," 38.

29 Ingersoll, "Abalone Fishery," 625; Todd J. Braje, *Shellfish for the Celestial Empire* (University of Utah Press, 2016), 97–134, 156–176. Braje uses archaeological evidence to contend that Chinese fishers did not deplete island abalone stocks.

30 Jordan, "Chinese Fishermen," 38–40; Ingersoll, "Abalone Fishery," 623; Collins, "Report," 31.

31 Ron Morgan, "The Impact of Racism on the Evolution of the Commercial Fishing Industry on Monterey Bay," master's thesis, California State University Dominguez Hills, 2001, 35.

32 McEvoy, "In Places," 6; Collins, "Report," 25; Connie Y. Chiang, *Shaping the Shoreline: Fisheries and Tourism on the Monterey Coast* (University of Washington Press, 2008), 1–37.

33 McEvoy, "In Places," 7–8; Braje, *Shellfish*, 156–176.

34 William A. Wilcox, "The Fisheries of the Pacific Coast," in *US Commission of Fish and Fisheries, Report of the Commissioner for the Year Ending June 30, 1893*, part 19 (GPO, 1895), 188–189.

35 "Sea Foam," *Los Angeles Times*, July 22, 1889.

36 "Rusticating on Catalina," *Los Angeles Times*, July 20, 1883.

37 Charles Frederick Holder, "An Isle of Summer: Santa Catalina Island," *Californian Illustrated Magazine* 3 (1892/1893): 67.

38 "With Rod and Reel," *Los Angeles Times*, January 1, 1899.

39 Holder, "Isle of Summer," 67.

40 Williamson, "The Haliotis," 22.

41 Charles Frederick Holder, *The Channel Islands of California* (A.C. McClurg, 1910), 29.

42 "Around Avalon," *Los Angeles Times*, June 30, 1889; "With Rod and Reel," *Los Angeles Times*; Lawrence Culver, *The Frontier of Leisure: Southern California and the Shaping of Modern America* (Oxford, 2010), 120.

43 "With Rod and Reel," *Los Angeles Times*.

44 Charles F. Holder, *Half Hours with the Lower Animals: Protozoans, Sponges, Corals, Shells Insects, and Crustaceans* (American Book Company, 1905), 109.

45 Culver, *Frontier of Leisure*, 90–91.

46 De Witt C. Lockwood, "Winter Sports on Catalina Island," *San Francisco Chronicle*, February 2, 1896.

47 Mrs. M. Burton Williamson, "Among the Shells: Their Study Opens an Interesting Field of Knowledge," *Los Angeles Times*, September 24, 1899; Liberty Hyde Bailey, *The Nature-Study Idea* (Doubleday, Page, 1903), 1–57.

48 Holder, "Isle of Summer," 67.

49 Holder, *The Channel Islands*, 303.

50 Culver, *Frontier of Leisure*, 89; Lockwood, "Winter Sports."

51 W. A. Baldwin, "The Abalone," *San Francisco Chronicle*, November 13, 1887.

52 "Sea Foam," *Los Angeles Times*, July 22, 1889; Lockwood, "Winter Sports"; "Catalina Island," *Los Angeles Times*, August 12, 1897; Annual Banquet, Tuna Club, Metropole Hotel, Santa Catalina Island, menu (1901), Rare Book Division, New York Public Library Digital Collections, http://menus.nypl.org/menus/15826.

53 "Hunting Otter and Abalones," *New York Times*, May 16, 1897.

54 Lockwood, "Winter Sports."

55 "Redondo Beach," *Los Angeles Times*, July 17, 1897.

56 "Observations: Exposure of the Deep-Laid Abalone Conspiracy, *Los Angeles Times*, August 31, 1905.

57 "Observations," *Los Angeles Times*.

58 Toshio Oba, "Gennosuke Kodani and the Abalone Fishermen from Chiba, Japan Who Introduced Diving Methods into the California Abalone Fishery," in *J.B. Phillips Historical Fisheries Report, Special Edition, A Project of the Monterey History and Art Association*, booklet, J.B. Phillips Collection, Monterey Public Library, California History Room; Lundy, *California Abalone Industry*, 15–17; Tim Thomas, *The Abalone King of Monterey: "Pop" Ernest Doelter, Pioneering Japanese Fishermen, and the Culinary Classic That Saved an Industry* (History Press, 2014), 17–22.

59 Minoru Nukada, "Historical Development of the Ama's Diving Activities," in *Physiology of Breath-Hold Diving and the Ama*, ed. Herman Rahn and Tetsuro Yokoyama, National Research Council, no. 1341 (National Academy of Sciences, 1965), 25–39; Ros Bandt, "Waiting for the Tide," track 1, *Isobue*, audio compilation that includes haiku, sound art, and oral history related to the *ama* (2008) (author collection).

60 Arne Kalland, *Fishing Villages in Tokugawa Japan* (University of Hawai'i Press, 1995), 163–164, 166–167.

61 Bandt, track 1, 33:00; Kalland, *Fishing Villages*, 166–168, 243.

62 Kumi Kato, "Waiting for the Tide, Tuning in the World: Traditional Knowledge, Environmental Ethics and Community," in *The Second International Small Island Cultures Conference*, Norfolk Island, 2006, ed. H. Johnson, 76–84; Bandt, track 1, 3:40, 33:00.

63 Naomi Hirahara, *Distinguished Asian American Business Leaders* (Greenwood Press, 2003), 104-106; Ruddle, *Administration*, intro. and chap. 1; Tsuyoshi Sasaki, "Direction of Fisheries (SUISUN) Education from a Historical Perspective in Japan," in *Fisheries and Aquaculture in the Modern World*, ed. Heimo Mikkola (IntechOpen, 2016), 160–161.

64 Sandy Lydon, *The Japanese in the Monterey Bay Region: A Brief History* (Capitola, 1997), 35.

65 Lydon, *The Japanese*, 35–37.

66 Lundy, *California Abalone Industry*, 26.

67 Kato, "Waiting," 79–80; Anne MacDonald, *Japan's 'Ama' Free Divers Keep Their Traditions* (2010) United Nations University, documentary. https://ourworld.unu.edu/en/japans-ama-free-divers-keep-their-traditions.

68 Lundy, *California Abalone Industry*, 24–25.

69 Lundy, *California Abalone Industry*, 18.

70 Williamson, "The Haliotis," 25.

71 Lundy, *California Abalone Industry*, 46–48.

72 William A. Wilcox, "Notes on the Fisheries of the Pacific Coast in 1899," Part 27, *US Commission of Fish and Fisheries Report for 1901* (GPO, 1902), 550; Lundy, *California Abalone Industry*, 19–20; "San Pedro: Japanese Abalone Divers," *Los Angeles Times*, November 15, 1899.

73 Wilcox, "Notes on the Fisheries," 550.

74 "Will Exterminate the Abalone," *Monterey New Era*, June 28, 1899.

75 "Will Exterminate," *Monterey New Era*.

76 "The Japanese and the Abalone," *Monterey New Era*, August 23, 1899.

77 "The Japanese," *Monterey New Era*.

78 Lydon, *The Japanese*, 84; Thomas, *Abalone King*, 28.

79 "Supervisors Will Protect the Abalone," *Monterey New Era*, October 4, 1899; "Monterey's Regulations: Abalone Cannot Be Fished for Out of Certain Limits," *Los Angeles Times*, October 7, 1899.

80 "The Development of the Southwest Crawfish [Lobsters]," *Los Angeles Times*, April 23, 1899; "Protecting the Abalone," *San Francisco Call*, October 7, 1902.

81 "To Preserve Shellfish," *Los Angeles Herald*, November 19, 1908.

82 Sam Ware, "The Abalones," *Overland Monthly*, December 1903, 534–536.

83 "Sportsmen Act as Supply of Abalone Is Endangered," *Fort Bragg Advocate and News*, August 17, 1927; Kitahara, quoted in Louis Martin, "Black Market Abalone," *Coast News*, nd, http://www.coastnews.com/ch17.htm.

84 "Abalones," *Seventeenth Biennial Report of the CBFC, 1901–1902*, 23–24. Commercial fishermen appealed the new state law, but it was upheld by a Santa Barbara court. For reference, the size limit for red abalone in Northern California's recently closed sportfishery was 7 inches.

85 "Abalones," *Eighteenth Biennial Report of the CBFC, 1903–1904*, 44; Williamson, "The Haliotis," 26. An amendment to the penal code in 1905 reduced the size limit for black abalones to 12 inches in circumference (about 4 inches in length). For reference, the state eventually established size limits of 5 inches and then 5¾ inches in length for black abalone. Modern sources say black abalone reach a maximum of 8 inches in length.

86 CDFG, Marine Region, *ARMP*, Appendix A, "Historical Summary of Laws and Regulations Governing the Abalone Fishery in California," A2, A13.

87 CDFG, *ARMP*, "Historical Summary," A8; Williamson, "The Haliotis," 26.

88 *Twenty-First Biennial Report of the CBFC, 1909–1910*, 40.

89 Lydon, *The Japanese*, 90.

90 CDFG, *ARMP*, "Historical Summary," A7.

91 Jack London, *Valley of the Moon* (McMillan, 1913/UC Press, 1999), 301.

92 Michael Orth, "Ideality to Reality: The Founding of Carmel," *California Historical Society Quarterly* 48, no. 3 (1969): 197–198; Franklin Walker, *The Seacoast of Bohemia* (Book Club of California, 1966/Peregrine Smith, 1973), 15–16.

93 Mary Austin, "George Sterling at Carmel," *American Mercury*, May 1927, 66.

94 Orth, "Ideality," 204. According to Austin, several of Jack London's characters were based on Sterling.

95 Mary Austin, "George Sterling," 66; Mary Austin, *Earth Horizon: Autobiography* (Houghton Mifflin, 1932), 298.

96 Austin, "George Sterling," 69; Austin, *Earth Horizon*, 299, 302.

97 Walker, *Seacoast*, 37.

98 Geoffrey Dunn, "Deep-Sea Matrimony: George Sterling and 'The Abalone Song,'" *Noticias del Puerto de Monterey: Bulletin of the Monterey History and Art Association* 57, no. 1 (2008): 3–4.

99 Otis Notman, "Visits to California Authors: Mary Austin, James Hopper, and George Sterling as They Appear in Their Homes by the Old Carmel Mission," *New York Times*, October 18, 1907.

100 "Hotbed of Soulful Culture, Vortex of Erotic Erudition; Carmel in California, Where Author and Artist Folk Are Establishing the Most Amazing Colony on Earth," *Los Angeles Times*, May 22, 1910.

101 Jack London, "The Valley of the Moon," *Cosmopolitan*, serialized, April–December 1913; London, *Valley*, 386–392.

102 *Twenty-First Biennial Report of the CBFC, 1909–1910*, 40.

103 *Twenty-Second Biennial Report of the CFGC, 1911–1912*, 10–11.

104 Charles Lincoln Edwards, "The Abalones of California," *Popular Science Monthly* 82 (June 1913): 534; Charles Lincoln Edwards, "The Abalone Industry in California," CFGC *Fish Bulletin* 1 (1913): 6.

105 Charles Lincoln Edwards, "Abalones of California," Smithsonian Institution, *Annual Report of the Board of Regents* (1913), 430.

106 Edwards, "Abalone Industry," 6–7; Callum Roberts, *The Unnatural History of the Sea* (Island Press, 2008), 260.

107 Edwards, "The Abalones of California," *Popular Science*, 549; Edwards, "Abalone Industry," 12.

108 Edwards, "Abalone Industry," 11.

109 Edwards, "Abalone Industry," 5. The California Board of Fish Commissioners changed its name to the Fish and Game Commission in 1910.

110 Edwards, "Abalone Industry," 12; Edwards, "Abalones of California," Smithsonian Institution *Annual Report*, 437; Edwards, "The Abalones of California," *Popular Science*, 559.

111 CDFG, *ARMP*, "Historical Summary," A8.

112 Lou Guernsey, "Another Blow at Japanese: Legislature Forbids Export of Abalone Shells," *Los Angeles Times*, April 17, 1913; Lou Guernsey, "Stops Clock to Sign Bills," *Los Angeles Times*, June 17, 1913.

113 Robert M. Sapolsky, *Behave: The Biology of Humans at Our Best and Worst* (Penguin Press, 2017).

114 Dunn, "Deep Sea Matrimony," 3–4.

115 Carl Doelter, "Charlie Chaplin Ate Here: Pop Ernest's Abalone and Seafood Restaurant: A Memoir by Carl Doelter," *Noticias de Monterey* 55, no. 4 (2006): 8; Thomas, *Abalone King*.

116 "My Trip to Santa Cruz," *Daily Alta California*, May 3, 1867, describes the Hotel Bromley serving abalone chowder in the 1860s; Arthur Inkersley, "Rare Shellfish," *Illustrated Scientific American* 93 (December 1905): 478.

117 "The Aulone," *Monterey Weekly Herald*; Brian D. Haley and Larry R. Wilcoxon, "Point Conception and the Chumash Land of the Dead," *Journal of California and Great Basin Anthropology* 21, no. 2 (1999): 222–225; "At Pacific Grove," *Pacific Rural Press*, May 28, 1892; "News and Business," *Los Angeles Times*, January 16, 1892; Williamson, "The Abalone," 23; John Dean Caton, "Abalone and Squid," *Forest and Stream* 36 (July 1891): 474; Thomas, *Abalone King*, 38–39; Les W. Field, *Abalone Tales* (Duke University Press, 2008), 64, 75.

118 Thomas, *Abalone King*, 38–40.

119 Pop Ernest business card, c. 1917–1919, from the collection of author Tim Thomas.

120 Hof Brau display ad, "Abalones," *San Francisco Chronicle*, July 24, 1914; Clarence Edgar Edwards, *Bohemian San Francisco: Its Restaurants and Their Most Famous Recipes; The Elegant Art of Dining* (P. Elder, 1914), 48.

121 "Of Interest to Fishermen at the Panama-Pacific International Exposition," *Forest and Stream* 83 (December 1914): 755.

122 Thomas, *Abalone King*, 58; Glenn A. Jones, "Quite the Choicest Protein Dish: The Costs of Consuming Seafood in American Restaurants, 1850–2006," in *Oceans Past: Management Insights from the History of Marine Animal Populations*, ed. David John Starkey and Michaela Barnard (Earthscan, 2008), 70. Jones says two million, but the NPS Panama-Pacific International Expo centennial website says the event hosted eighteen million visitors; http://www.nps.gov/goga/learn/historyculture/ppie.htm.

123 Lydon, *The Japanese*, 87.

124 Thomas, *Abalone King*, 46, 60–61, 68, 80; Doelter, "Charlie Chaplin Ate Here," 13–16.

125 Doelter, "Charlie Chaplin Ate Here," 17; Thomas, *Abalone King*, 104.

126 Louis J. Stellman, "Interesting Westerners: Ernest Doelter of Monterey Knows More about the Abalone than Any Other Chef in California," *Sunset*, April 1920, 49.

127 Thomas, *Abalone King*, 62.

CHAPTER 5

1 William W. Curtner, "Observations of the Growth and Habits of the Red and Black Abalone," master's thesis, Stanford University, May 1917; Paul Bonnot, "California Abalones," *California Fish and Game* 26, no. 3 (1940): 206. Abalone size limits were measured in circumference until 1917; CDFG, Marine Region, *ARMP*, Appendix A, "Historical Summary of Laws and Regulations Governing the Abalone Fishery in California," A4, A14.

2 Harold Heath, "The Abalone Question," typewritten copy of article appearing in *Monterey American* on March 15, 1917, and in the *Monterey Daily Cypress* on March 16, 1917, dated June 11, 1925, CSA, Natural Resources, Fish and Game Division, Marine Fisheries, Abalone, 1916–1953, F3735:640; Harold Heath, "California Clams," *California Fish and Game* 2, no. 4 (1916): 175–177.

3 Arthur F. McEvoy, *The Fisherman's Problem: Ecology and Law in California Fisheries, 1850–1980* (Cambridge University Press, 1986), 158–159; "N.B. Retires," *California Conservationist* (November 1939): 5; Frances N. Clark, "California Marine Fisheries Investigations, 1914–1939," *CalCOFI Report* 23 (1982): 25–28.

4 CDFG, *ARMP*, "Historical Summary," A10; M. L. Church, "California Association Meeting," *Forest and Stream* 75 (October 1910): 697.

5 W. R. Holman to George D. Northenholt (director of natural resources), letter, June 8, 1935, CSA, Natural Resources, Administration, Director's Subject File, Abalone-Deer, F3735:108.

6 John Woolfenden, "One Man's Fight to Save the Abalone," *Monterey Herald*, June 13, 1976; "Further Discussion of Abalone Question—More Reasons for Protection," *Monterey Daily Cypress*, March 7, 1917.

7 Ann Vileisis, *Discovering the Unknown Landscape: A History of America's Wetlands* (Island Press, 1997), 151–155; Church, "California Association Meeting," 697.

8 "W. R. Holman Would Protect Abalones," *Monterey Daily Cypress*, February 8, 1917; "Text of Martin's Abalone Bill," *Monterey Daily Cypress*, March 19, 1917; Holman to Northenholt, letter, 1935; "Carmel Colony Wants Law to Shield Abalone," *San Francisco Chronicle*, January 11, 1917.

9 *Thirty-Fifth Biennial Report of the CFGC, 1936–37*, 60.

10 N. B. Scofield to F. G. Knight, letter, June 11, 1925, CSA, Natural Resources, Fish and Game Division, Marine Fisheries, Abalone, 1916–1953.

11 Harold Heath, "The Abalone Question," *Monterey Daily Cypress*, March 16, 1917. According to Holman, the article was printed first in the *Pacific Grove News* on September 18, 1916, and then reprinted in many newspapers. The article was later reprinted in *California Fish and Game* 11, no. 3 (1925): 138–139.

12 Heath, "Abalone Question."

13 Heath, "Abalone Question"; N. B. Scofield, "Conservation Laws Provide Ample Protection for Abalones," *California Fish and Game* 16, no. 1 (1930):13–15.

14 Heath "Abalone Question."

15 An Abalone Fan, "Diving for Abalones Should Be Prohibited, Is Contention," *Monterey Daily Cypress*, March 17, 1917.

16 "Abalone Question from Another Point of View," *Monterey Daily Cypress*, March 4, 1917.

17 "Abalone Question from Another Point of View."

18 "Abalone Question from Another Point of View."

19 Woolfenden, "One Man's Fight," 10; "Further Discussion of Abalone Question"; H. A. Greene, "Abalone Controversy," *Monterey Daily Cypress*, March 17, 1917; W. W. Curtner, "Present Law Amply Protects the Abalone," *Monterey Daily Cypress*, March 16, 1917; W. R. Holman, "The Question of Further Protecting the Abalone," *Monterey Daily Cypress*, March 18, 1917.

20 Harold Heath, "California Clams," 175–178; "Conflicting Views of Abalone Question at Public Hearing," *Monterey Daily Cypress*, April 1, 1917.

21 Holman to Northenholt, letter, 1935; Woolfenden, "One Man's Fight," 10.

22 "Abalone Bill Passes Assembly," *Monterey Daily Cypress*, March 28, 1917; "Grove Civic Club Proposes Compromise of Abalone Question," *Monterey Daily Cypress*, April 4, 1917.

23 "Compromise on Abalone Bills," *Monterey Daily Cypress*, April 12, 1917; N. B. Scofield, "Commercial Fishery Notes," *California Fish and Game* 7, no. 3 (1921): 174.

24 Scofield, "Conservation Laws," 13.

25 "State Official Says Abalone Are Plentiful: N.B. Scofield Scoffs at Rumor Shellfish May Become Extinct," *San Francisco Chronicle*, February 15, 1920.

26 "State Official Says," *San Francisco Chronicle*, 35.

27 Syl MacDowell, "Abalone Is Saved from Extinction," *Los Angeles Times*, February 11, 1923.

28 N. B. Scofield to Bureau of Foreign and Domestic Commerce, letter, March 10, 1926, CSA, Natural Resources, Administration, Director's Subject File, Abalone-Deer.

29 CDFG, *ARMP*, "Historical Summary," A14.

30 "Residents Protest against Abalone Divers," *Mendocino Beacon*, August 2, 1927.

31 N. B. Scofield to Charles R. Perkins of Fort Bragg, letter, regarding the commercial taking of abalone, August 16, 1927, CSA, Natural Resources, Fish and Game Division Marine Fisheries, Abalone, 1916–1953; N. B. Scofield, "Abalone Safe from Extermination," *California Fish and Game* 14, no. 1 (1928): 87.

32 "Residents Protest," *Mendocino Beacon*.

33 "Many Club Members Hear Col. Schofield [*sic*] Talk on Abalonies," *Fort Bragg Advocate and News*, May 18, 1932; N. B. Scofield to Mr. Farley, letter, May 28, 1932, regarding abalones in Mendocino County, CSA, Natural Resources, Fish and Game Division, Marine Fisheries, Abalone, 1916–1953.

34 CDFG, *ARMP*, "Historical Summary," A10; "Monterey Losing Prestige as 'Abalone Capital' due to Opening Southern Waters," *Peninsula Daily Herald*, April 2, 1932; "Extinction of Abalones Not Feared by Scofield," *Peninsula Daily Herald*, September 13, 1929; "Scofield Says Abalones Not Being Depleted," *Herald Recorder* (Arroyo Grande), September 19, 1929, clipping; T. W. Shilling (Pismo Beach) to Capt. C. H. Groat (Terminal Island), memo, April 20, 1931; correspondence about shipwreck and violations of Japanese diver T. Sumi, May 8, 1931; handwritten note from H. Bluett, Deputy of Patrol Dept., April 22, 1931, last four items from CSA, Natural Resources Agency, Fish and Game Division Marine Fisheries, Abalone, 1916–1953; "Drastic Law Is Advocated for Abalone Loss," *Monterey Trader*, February 23, 1933.

35 "Local Abalone Fleet Prepares for Operation," *Monterey Trader*, March 9, 1933.

36 CDFG, *ARMP*, "Historical Summary," A10.

37 Richard S. Croker, "Abalones," *Fish Bulletin* 30 (1930): 65–66.

38 "Don't Kidnap Baby Abalones," *Los Angeles Times*, November 24, 1922; Croker, "Abalones," 65–66; CDFG, *ARMP*, "Historical Summary," A1.

39 A. L. "Scrap" Lundy, *The California Abalone Industry: A Pictorial History* (Best, 1997), 55–56.

40 Lundy, *California Abalone Industry*, 70.

41 Lundy, *California Abalone Industry*, 61.

42 Lundy, *California Abalone Industry*, 75.

43 Lundy, *California Abalone Industry*, 56.

44 Lundy, *California Abalone Industry*, 66.

45 Lundy, *California Abalone Industry*, 67.

46 Lundy, *California Abalone Industry*, 66–69, 96–97, 105.

47 Lundy, *California Abalone Industry*, passim.

48 Lundy, *California Abalone Industry*, 76.

49 Lundy, *California Abalone Industry*, 80.

50 Lundy, *California Abalone Industry*, 100.

51 Glen Bickford, "A Midwesterner Reports from the Pacific Floor, 1936," in Dorothy L. Gates and Jane H. Bailey, *Morro Bay's Yesterdays*, (El Moro Publications, 1993 [1982]), excerpt at http://www.historyinslocounty.org/Glen%20Bickford.htm.

52 Lundy, *California Abalone Industry*, 67, 109.

53 J. B. Phillips, "Abalone," *Fish Bulletin* 49 (1936): 107–109; Lundy, *California Abalone Industry*, 67; "Monterey," in the commercial fisheries trade journal *California Fisheries* 1, no. 1 (1928): 10; "Drastic Law," *Monterey Trader*.

54 J. B. Phillips to N. B. Scofield, letter, Abalones and Coast Highway Construction, March 29, 1933; J. B. Phillips, "Effects of Carmel-San Simeon Road Construction on Abalone & Kelp," California State Fisheries Lab, March 25, 1933, both in CSA, Natural Resources Agency, Fish and Game Division, Marine Fisheries, Abalone, 1916–1953.

55 NOAA, Earth System Research Laboratory, Physical Sciences Division, "Top 24 Strongest El Niño and La Niña Event Years by Season," https://www.esrl.noaa.gov/psd/enso/climaterisks/years/top24enso.html, accessed August 15, 2019.

56 "Drastic Law," *Monterey Trader*.

57 "Monterey Losing Prestige," *Peninsula Daily Herald*; "Local Abalone Fleet," *Monterey Trader*.

58 Lundy, *California Abalone Industry*, 83–84.

59 Thomas D. Reviea, "Practical Conservation," newspaper clippings, part 1 and part 3, nd, CSA, Natural Resources, Fish and Game Division, Marine Fisheries, Abalone, 1916–1953.

60 Thomas D. Reviea to Paul Bonnot, letter, March 10, 1939, CSA, Natural Resources, Fish and Game Division, Abalone, 1916–1953.

61 Paul Bonnot, "Abalones in California," *California Fish and Game* 16, no. 1 (1930): 3–11.

62 Paul Bonnot, "California Abalones," *California Fish and Game* 26, no. 3 (1940): 208; Paul Bonnot, "The Score on the California Abalones," *California Conservationist* (February 1941): 4.

63 Bonnot, "California Abalones," 209; "Abalones: Survey Is Continued," *California Conservationist* (November 1939): 5; Bonnot, "The Score," 5–8.

64 Bonnot, "The Score," 8.

65 "Abalones: Search Goes Deeper," *California Conservationist* (September 1939): 5.

66 Bonnot, "California Abalones," 208–209.

67 McEvoy, *Fisherman's Problem*, 261.

68 CDFG, *ARMP*, "Historical Summary," A3.

69 Paul Bonnot to N. B. Scofield, letter, December 14, 1935, CSA, Natural Resources, Fish and Game Division, Marine Fisheries-Abalone, 1916–1953; Richard Van Cleve, "Report of the Bureau of Marine Fisheries," *Thirty-Seventh Biennial Report of the CFGC, 1942–43*, 53.

70 "Abalones: Search Goes Deeper," *California Conservationist*, 5; "N.B. Retires," *California Conservationist*, 5, 17.

71 "Abalones: Search Goes Deeper," *California Conservationist*, 5.

72 "Abalones: Search Goes Deeper," *California Conservationist*, 5; Van Cleve, "Report," 53.

73 Connie Y. Chiang, *Shaping the Shoreline: Fisheries and Tourism on the Monterey Coast* (University of Washington Press, 2008), 106–108; Lundy, *California Abalone Industry*, 45–46, 90–91; Paul Bonnot, "The Abalones of California," *California Fish and Game* 34 (1948): 141–169.

74 Richard Van Cleve, "Report of the Bureau of Marine Fisheries," *Thirty-Eighth Biennial Report of the CFGC, 1942–1943*, 41; Lundy, *California Abalone Industry*, 90–91,108, 113.

75 Tim Thomas, Roy Hattori, and Art Seavey, Abalone history presentation, March 28, 2011, Clement Monterey Hotel; Roy Hattori, interview by Tim Thomas and Sandy Lydon, audio recording, nd, from the collection of author Tim Thomas. Mr. Hattori died later that year.

76 Paul Bartsch, "The West American Haliotis," *Proceedings of the US National Museum* 89, no. 3094 (1940): 49–58. Note the complete species name is *Haliotis sorenseni*, Bartsch 1940.

77 Tim Thomas, interview, March 27, 2011.

78 Roy Hattori, interview by Thomas and Lydon.

CHAPTER 6

1 Coles Phinizy, "The Old Men of the Sea," *Sports Illustrated*, August 23, 1965.

2 Jack Prodanovich and Wally Potts, interview by Craig Carter, August 29, 1983, San Diego Historical Society, Oral History Program; Gordon Smith, "Action: The Bottom Scratchers," *Los Angeles Times*, November 13, 1986.

3 Terry Maas, "After 72 Years, the Famed San Diego Bottom Scratchers Club Has Closed Its Doors," BlueWater Freedivers blog (2005), http://www.freedive.net/bottom_scratchers/bottom_scratchers.htm.

4 Eric Hanauer, *Diving Pioneers: An Oral History of Diving in America* (Watersport, 1994), 26; Phinizy, "Old Men," 78.

5 Smith, "Action."

6 Smith, "Action"; Maas, "After 72 Years."

7 Smith, "Action"; David Hellyer, "Goggle Fishing in California Waters," *National Geographic*, May 1949.

8 Hellyer, "Goggle Fishing," 615–631.

9 Maas, "After 72 Years."

10 Hanauer, *Diving Pioneers*, 74–76, 53.
11 Hanauer, *Diving Pioneers*, 21; the dive club names come from mentions in *Skin Diver* magazine through 1950s.
12 "Recipe for Abalone Chowder," *Skin Diver*, December 1951; "Recipe for Abalone Patties," *Skin Diver*, February 1952. The recipes were eventually dropped from the magazine.
13 "California's Low Tide Harvest ... the Precious Abalone," *Sunset*, May 1953; "Try Abalone in a Stew or Chowder," *Sunset*, November 1958.
14 Prodanovich and Potts, interview by Carter, 1983; Maas, "After 72 Years."
15 Prodanovich and Potts, interview by Carter, 1983; CDFG, Marine Region, *ARMP*, Appendix A, "Historical Summary of Laws and Regulations Governing the Abalone Fishery in California," A3.
16 Smith, "Action."
17 CDFG, *ARMP*, "Historical Summary," A3, A12.
18 Keith W. Cox, "California Abalones, Family Haliotidae," *Fish Bulletin* 118 (1962): 80.
19 Woody Dimel, "Los Angeles Neptunes," *Skin Diver*, February 1952.
20 Transcript of Hearing, October 2, 1957, CSA, Assembly Committee on Fish and Game, 94–105, LP163:271-272; Bob Kirby, *Hard Hat Divers Wear Dresses* (Olive Press Publications, 2002), 12–13.
21 CDFG, *ARMP*, "Historical Summary," A3.
22 "Council of Diving Clubs," *Skin Diver*, June 1953.
23 Bill Barada, "CA Council of Diving Clubs," *Skin Diver*, February 1954.
24 Smith, "Action."
25 *Forty-First Biennial Report of the CDFG, 1948–1950*, 56.
26 Tom White, "Tire Irons in Angling," *New York Times*, May 11, 1941.
27 Cox, "California Abalones," 98.
28 *Forty-Second Biennial Report of the CDFG, 1950 to 1952*, 48–49.
29 CDFG, *ARMP*, "Historical Summary," A3.
30 Keith Cox, "A Preliminary Report of the Investigation of the California Abalone Resource Being Conducted by the California Department of Fish and Game on Direction of the Legislature," 1955, 2, CSL; E. C. Greenhead (biostatistical supervisor) to Phil M. Roedel (marine resources manager), Meeting with the Abalone Industry, Santa Barbara, December 16, 1960; intraoffice correspondence, December 23, 1960, CSA, Fish and Game, Deputy Director's Files, Marine Resources Branch, F3498:545; Jim Thomas, "Abalone Tagging Program," AP Newsfeatures, manuscript, October 29, 1951, CSA, Natural Resources, Fish and Game Division, Marine Fisheries, Abalone, 1916–1953.
31 Thomas, "Abalone Tagging Program."
32 Cox, "California Abalones," 5.
33 Cox, "Preliminary Report," 3.
34 Cox, "California Abalones," 6; Kirby, *Hard Hat Divers*, 15–16.
35 "Abalone Survey Yields Interesting Results off of Peninsula," *Monterey Peninsula Herald*, June 19, 1957, clipping in Abalone Study, 1957 envelope no. 10, Monterey Public Library, California History Room.
36 Cox, "Preliminary Report," 5.
37 "Report Cites Reasons Why North Coast Waters Cannot Support Commercial Abalone Fishery," *Outdoor California*, 18, no. 5 (May 1957): 1, 3, 10; Cox, "California Abalones," 112.
38 "Report Cites Reasons," *Outdoor California*, 10; Jerald S. Ault and John D. DeMartini, "Movement and Dispersion of Red Abalone, *Haliotis rufescens*, *California Department of Fish and Game* 73, no. 4 (1987): 196–213.
39 Cox, "California Abalones," 104.
40 "Report," *Outdoor California*, 3; Cox, "Preliminary Report," 9; Keith W. Cox, "Notes on the California Abalone Fishery," in *Proceedings of the National Shellfisheries Association* 48 (1957): 105; Transcript of Hearing, October 2, 1957, 94.
41 *Forty-Fourth Biennial Report of the CDFG, 1954–1956*, 50.
42 Keith Cox, CDFG Cruise Reports, 62-S5, 62S6, 62M2, August 15–27 and September 5–17, 1962, Abalone; Cox, "California Abalones," 80, 87; *Forty-Fifth Biennial Report of the*

CDFG, 1956–1958, 66–67; CDFG, *ARMP*, "Historical Summary," A10; Keith W. Cox, "Notes," 105.

43 Parke H. Young, "Some Effects of Sewer Effluent on Marine Life," *California Fish and Game* 50 (1964): 33–41; A. M. Rawn, "Narrative–C.S.D." (County Sanitation Districts of Los Angeles County, 1965), 5, www.sewerhistory.org/sewerarticles/california/, accessed April 29, 2018.

44 Terence Kehoe and Charles Jacobson, "Environmental Decision Making and DDT Production at Montrose Chemical Corporation of California," *Enterprise and Society* 4, no. 4 (2003): 665.

45 Cox, "California Abalones," 80–83; *Forty-Fifth Biennial Report of the CDFG, 1956–1958*, 66.

46 Buzz Owen, interview, September 19, 2019; Cox "California Abalones," 52; CDFG, *ARMP*, "Historical Summary," A14.

47 Cox, "California Abalones," 52

48 *Forty-Fifth Biennial Report of the CDFG*, 66.

49 CDFG, abalone landings data, 1958–1968, Ian Taniguchi, correspondence, September 4, 2019.

50 Mia J. Tegner et al., "Climate Variability, Kelp Forests, and the Southern California Red Abalone Fishery," *Journal of Shellfish Research* 20, no. 2 (2001): 755–763.

51 *Forty-Seventh Biennial Report of the CDFG, 1960–1962*, 32; A. L. "Scrap" Lundy, *The California Abalone Industry: A Pictorial History* (Best, 1997), 156; Tena Bettencourt, interview, November 22, 2010.

52 Lundy, *California Abalone Industry*, 112.

53 Kirby, *Hard Hat Divers*, 73–74, 82–83.

54 *Forty-Seventh Biennial Report of the CDFG*, 32.

55 Cox, "California Abalones," 26, 50, 88, 98–99, 102, 118.

56 *Forty-Seventh Biennial Report of the CDFG*, 32.

CHAPTER 7

1 Richard A. Boolootian, "The Distribution of the California Sea Otter," *California Fish and Game* 47, no. 3 (1961): 287; Richard Ravalli, "The Near Extinction and Reemergence of the Pacific Sea Otter, 1850–1938," *Pacific Northwest Quarterly* 100, no. 4 (2009): 188.

2 Paul W. Wild and Jack A. Ames, "A Report on the Sea Otter, *Enhydra lutris L.*, in California," CDFG Marine Resources Technical Report 20 (1974), 8, CSL.

3 Rolf L. Bolin, "Reappearance of the Southern Sea Otter along the California Coast," *Journal of Mammalogy* 19, no. 3 (1938): 301; Augustin S. MacDonald, "The Sea Otter Returns to the California Coast," *California Historical Society Quarterly* 17, no. 3 (1938): 243. In a newspaper article and unpublished account by Mr. Howard Sharpe, he takes full credit for the discovery and adds greater detail about the initial doubt by experts. Howard Granville Sharpe, "The Discovery of the 'Extinct' Sea Otter," manuscript at http://www.seaotters.org/pdfs/extinct.pdf.

4 "Sea Otter Herd Is Sighted: Scientists in Dither," *Monterey Peninsula Herald*, March 28, 1938.

5 MacDonald, "Sea Otter Returns," 243.

6 Edna Fisher, "Habits of the Southern Sea Otter," *Journal of Mammalogy* 20, no. 1 (1939): 22; Senate Fact Finding Committee on Natural Resources, Subcommittee on Sea Otters, "Affect [*sic*] of the Sea Otter on the Abalone Resource," transcript of hearing, San Luis Obispo (November 19, 1963), 21, CSL; Roy Hattori, interview by Tim Thomas and Sandy Lydon, audio recording, nd, from collection of author Tim Thomas.

7 Senate Fact Finding Subcommittee on Sea Otters, 21.

8 Statement of Mr. Ernest C. Porter, Making a Limited Appearance on Behalf of the Abalone Industry, US Atomic Energy Commission, Hearing on Pacific Gas and Electric Company's Proposed Diablo Canyon Nuclear Power Plan, transcript of proceedings (February 20–21, 1968), 2, in Moss Landing Marine Labs Library, Archives, California Sea Otter Controversy, miscellaneous papers.

9 CDFG, Operations Research Branch, "A Proposal for Sea Otter Protection and Research and Request for the Return of Management to the State of California," draft, vol. 1 (January 1976), 187, CSL.

10 Wild and Ames, "Report," 19.

11 Harold Bissell and Frank Hubbard, "Report on the Sea Otter, Abalone and Kelp Resources in San Luis Obispo and Monterey Counties and Proposals for Reducing the Conflict between the Commercial Abalone Industry and the Sea Otter, Requested by Senate Concurrent Resolution 74 in the 1967 Legislative Session," CDFG (January 1968), 17, CSL.

12 Sachiye Kuramitsu, "Sea Otters Pose Threat to Yield of Abalone," Los Angeles Times, October 6, 1963.

13 Keith Cox, CDFG Cruise Reports, 63-S-5, 63-M-1, August 1–16 and September 24, 1963, Abalone.

14 Kuramitsu, "Sea Otters."

15 "State Will Continue Protection of Sea Otters," Los Angeles Times, October 17, 1963.

16 Dewey Linze, "Conservationists Rise to Sea Otters' Defense: UCLA Zoologist Cites Treaty Protection, Says Abalone Comprise Only 5 percent of Diet," Los Angeles Times, November 3, 1963.

17 John H. Prescott, "The Letter Page: Sea Otter Wins Ally in War of the Abalone," Los Angeles Times, October 26, 1963; Tom Harris, "Abalone Hunters Must Take Some Blame for Depletion," San Luis Obispo County Telegram-Tribune, August 16, 1963.

18 Senate Fact Finding Subcommittee on Sea Otters, 55; "Sea Otter-Abalone Controversy," transcript of conference, Moss Landing Marine Laboratories, November 24, 1969, 22–27, CSL.

19 Senate Fact Finding, 6.

20 Senate Fact Finding, 42.

21 Senate Fact Finding, 7.

22 Keith W. Cox, "California Abalones, Family Haliotidae," Fish Bulletin 118 (1962): 80.

23 Senate Fact Finding, 8.

24 Senate Fact Finding, 27, 63, 71.

25 Senate Fact Finding, 59, 110, 136.

26 Senate Fact Finding, 138.

27 CDFG, "Sea Otter and Its Effect on the Abalone Resource," report, nd (c. 1967), 10, CSL.

28 Ernie Porter, "Letter to the Sportsmen," June 6, 1967, and Gloria V. Pierce, "Letter to the Sportsmen," June 7, 1965 [sic], in "Abalone vs. Sea Otters," packet submitted to CDFG director Walter Shannon by Members of the Abalone Industry, Moss Landing Marine Labs Library, Archives, California Sea Otter Controversy, Miscellaneous Papers, 1974.

29 "Sea Otter-Abalone Controversy," 88.

30 Bissell and Hubbard, "Report," 21–24.

31 Lynn Lilliston, "Conservationists Fighting Drive against Sea Otters," Los Angeles Times, March 30, 1970; Wolfgang Saxon, "Margaret Wentworth Owings, 85, Defender of Wild Creatures," New York Times, January 31, 1999; "Guide to Margaret Wentworth Owings Papers," UC Berkeley, Bancroft Library.

32 Margaret Owings, Friends of the Sea Otter newsletter, First Issue, Otter Raft 1 (June 1969): 1; see also nos. 2–4, https://www.seaotters.org/newsletters.

33 Lilliston, "Conservationists."

34 "Sea Otter-Abalone Controversy," 74–78.

35 Wild and Ames, "Report," 14.

36 "Sea Otter-Abalone Controversy," 23–27.

37 "Sea Otter-Abalone Controversy," 82.

38 "Sea Otter-Abalone Controversy," 78; Senate Fact Finding, 29–30.

39 Harris, "Abalone Hunters," 1.

40 Earl E. Ebert, "A Food Habits Study of the Southern Sea Otter, Enhydra lutris neries," California Fish and Game 54 no. 1 (1968): 37.

41 Ebert, "Food Habits Study," 35.

42 Ebert, "Food Habits Study," 39–40.

43 Ebert, "Food Habits Study," 40; Pioneering scuba researcher James McLean had surveyed sea otter habitat near Granite Creek in 1959–1960 and first noted the lack of red urchins, abalones constrained to cracks, and rich algal communities that flourished without the grazing pressures of urchins. James H. McLean, "Sublittoral Ecology of Kelp Beds," *Biological Bulletin* 122, no. 1 (1962): 101–103.

44 Earl Ebert, interview, April 3, 2013.

45 Ebert, interview, 2013.

46 John Radovich, "Relationship of Some Marine Organisms of the Northeast Pacific to Water Temperatures Particularly during 1957–1959," *Fish Bulletin* 112 (1961): 5–60; Wheeler North, interview by Shelley Erwin, California Institute of Technology Oral History Project (2001), 59–62, http://oralhistories.library.caltech.edu/34.

47 Wheeler J. North and John S. Pearse, "Sea Urchin Population Explosion in Southern California Coastal Waters," *Science*, new series 167, no. 3915 (1970): 209.

48 North and Pearse, "Sea Urchin," 209.

49 James A. Estes and John F. Palmisano, "Sea Otters: Their Role in Structuring Nearshore Communities," *Science* 185, no. 4156 (1974): 1058–1060.

50 Robert T. Paine, "A Note on Trophic Complexity and Community Stability," *American Naturalist* 103, no. 929 (1969): 91–93; James A. Estes, *Serendipity: An Ecologist's Quest to Understand Nature* (University of California Press, 2016), 28–44.

51 Estes and Palmisano, "Sea Otters," 1060.

52 Robert Steneck et al., "Kelp Forest Ecosystems: Biodiversity, Stability, Resilience and Future," *Environmental Conservation* 29, no. 4 (2002): 448.

53 Wild and Ames, "Report," 58; "Capital Standstill on Otter Question," *Los Angeles Times*, April 9, 1970.

54 Lilliston, "Conservationists."

55 Lilliston, "Conservationists."

56 Earl E. Ebert, "The Sea Otter in California's Wildlife," *Transactions of the Western Section of the Wildlife Society* 4 (1968): 20; Ebert, interview, 2013.

57 Margaret Owings, "The Outlook," *Otter Raft* 5 (May 1971).

58 Lilliston, "Conservationists."

59 Lilliston, "Conservationists."

60 "Capital Standstill," *Los Angeles Times*.

61 Gloria V. Pierce, "Sea Otter Curb Sought," *Los Angeles Times*, April 13, 1970. This quote is from a letter to the editor printed shortly after the hearing; I could find no transcript from the actual hearing.

62 "Capital Standstill," *Los Angeles Times*.

63 "Capital Standstill," *Los Angeles Times*.

64 Lynn Lilliston, "Dropping of Otter Control Bill Hailed: Schoolchildrens' Letters to Legislator Called Factor," *Los Angeles Times*, May 8, 1970.

65 Margaret Owings, "The Outlook," *Otter Raft* 4 (December 1970).

66 Wild and Ames, "Report," 69.

67 Jane Eshleman Conant, "Our Sea Otters Thrive Again: Shellfish Are Scurrying," *San Francisco Sunday Examiner and Chronicle*, June 13, 1971, clipping in CSA, CDFG, Director's Files, Marine Resources Branch, Sea Otters, 1964, F3498:372.

68 Conant, "Our Sea Otters."

69 Jeffrey St. Clair, "30 Years After: The Legacy of America's Largest Nuclear Test," *In These Times*, August 8, 1999, Institute for Public Affairs blog, http://inthesetimes.com/project-censored/stclair2317new.html; Dean W. Kohlhoff, *Amchitka and the Bomb* (University of Washington Press, 2002), 53, 110.

70 "The Unsinkable Sea Otter," *The Undersea World of Jacques Cousteau*, TV documentary (1971).

71 "Senate OKs Ban on Taking Sea Mammals, " *Los Angeles Times*, July 27, 1972; "Determination That the Southern Sea Otter Is a Threatened Species," *Federal Register*, January 14, 1977, 2965–2968.

CHAPTER 8

1 Tom Wolfe, *The Pumphouse Gang* (Farrar, Straus and Giroux, 1968), 19–39: Gary Wickham, interview, March 28, 2011.

2 Wickham, interview, 2011; for more on the Mac Meda Destruction Company, see, http://macmedadestruction.com/la-jolla-abalone-dive-point-loma, accessed March 19, 2015.

3 "Abalone Season Opens March 16," *Sunset,* March 1968, 55.

4 "Abalone Chowder at the Beach: First You Must Get Your Abalone," *Sunset,* October 1972, 196, 198, 200.

5 A. L. "Scrap" Lundy, *The California Abalone Industry: A Pictorial History* (Best, 1997), 158–159.

6 Konstantin A. Karpov et al., "Serial Depletion and the Collapse of the California Abalone (*Haliotis* spp.) Fishery," in *Workshop on Rebuilding Abalone Stocks in British Columbia,* ed. Alan Campbell, Canadian Special Publication in Fisheries and Aquatic Science 130 (NRC Research Press, 2000), 11–24.

7 Lundy, *California Abalone Industry,* 158–159, 160, 179; Biliana Cicin-Sain et al., "Management Approaches for Marine Fisheries: The Case of the California Abalone," UC Sea Grant College Program Publication 54 (January 1977): 4–5; Karpov et al., "Serial Depletion," 11, 19; CDFG, *ARMP,* "Historical Summary of Laws and Regulations Governing the Abalone Fishery in California," Appendix A16.

8 Pete Haaker, interview, September 10, 2019; Lad Handelman, interview, March 28, 2014.

9 Richard Burge, Steven Schultz, and Melvyn Odemar, "Draft Report on Recent Abalone Research in California with Recommendations for Management," CDFG, Operations Research Branch and Marine Resources Region, California Resources Agency, presented to the CFGC, January 17, 1975, San Diego, 1.

10 Burge et al., "Draft Report," 1.

11 Richard Burge, interview, January 5, 2011.

12 Burge, interview, 2011; Ralph Young, "Bringing Back the Abalone," *Outdoor California* 36, no. 4 (1974), 26; Burge et al., "Draft Report," 2.

13 Melvyn Odemar, interview, December 29, 2010; Richard Burge, CDFG Cruise Report 73-KB-17, July 18–26, 1973, Abalone; Richard Burge, CDFG Cruise Report, 73-KB-18, August 7–11, 1973, Abalone.

14 Dave Zeiner, "Saving a Valuable but Limited Resource—California's Abalone," *Outdoor California* 37, no. 4 (1975): 9.

15 Zeiner, "Saving," 9.

16 Zeiner, "Saving," 9.

17 Burge et al., "Draft Report," passim; Alan J. Wyner et al., "Politics and Management of the California Abalone Fishery," *Marine Policy* (October 1977): 327.

18 Burge et al., "Draft Report," 2–3; Dan Miller, "Skindivers, Abalone, and Sea Otters," *Outdoor California* 36, no. 4 (1974): 1.

19 Burge et al., "Draft Report," 3–4.

20 Burge et al., "Draft Report," 4; Richard Burge, CDFG Cruise Reports, 74-KB-8, 74-M-1, March 6–12, 1974, Abalone-Lobster Investigations; Burge, interview, 2011; Burge correspondence, May 29, 2017.

21 Richard Burge, CDFG Cruise Reports, 74-KB-8, 74-M-1, March 6–12, 1974, Abalone-Lobster Investigations; Burge et al., "Draft Report," 4, 16–17.

22 Burge et al., "Draft Report," 3–4; Cicin-Sain et al., "Management Approaches," 8.

23 Burge et al., "Draft Report," 4.

24 Burge et al., "Draft Report," 5–8.

25 Burge et al., "Draft Report," 8.

26 Burge et al., "Draft Report," 12–13.

27 Burge et al., "Draft Report," 9–10.

28 Burge et al., "Draft Report," 10–11.

29 Burge et al., "Draft Report," 10–11.

30 Burge et al., "Draft Report," 16–18.

31 Wyner et al., "Politics," 328; CFGC, meeting minutes, January 17, 1975, 25-27; CFGC, meeting minutes, October 3, 1975, 11, both CSA.

32 Earl Gustkey, "The Southland's Disappearing Abalone: State Acts to Increase Supply of Once Plentiful Delicacy in Local Waters," *Los Angeles Times*, November 8, 1977; Burge, interview, 2011.

33 CFGC, meeting minutes, October 3, 1975, 11 (Burge quote); Wyner et al., "Politics," 335; Cicin-Sain et al., "Management Approaches," 94.

34 CFGC, meeting minutes, October 3, 1975, 12; Wyner et al., "Politics," 329; Cicin-Sain et al., "Management Approaches," 94–95.

35 Richard A. Burge, "Abalone Management Report," Presentation to the California Fish and Game Commission, October 3, 1975, 3–4, CSA, Bill File AB 2224, Chapter 327, MF3:3 (36); Wyner et al., "Politics," 336–337; Cicin-Sain et al., "Management Approaches," 114.

36 Burge et al., "Draft Report," 5.

37 Odemar, interview, 2010; CDFG, *ARMP*, "Historical Summary," A2–A5; Zeiner, "Saving," 9; Cicin-Sain et al., "Management Approaches," 93.

38 Burge, "Abalone Management," 7.

39 Cicin-Sain et al., "Management Approaches," 95–96; Burge, "Abalone Management," 4; CFGC, meeting minutes, October 3, 1975, 12.

40 Wyner et al., "Politics," 329; Bill File AB 2224, Third Reading, January 6, 1976, CSA, MF8:4 (66).

41 Ivan P. Colburn, "The Decline of Abalone Fishing," *Los Angeles Times*, November 23, 1977.

42 "110 Years of Fish and Game History," *Outdoor California* (1980): 4.

43 Burge, interview, 2011; Mia Tegner, "An Analysis of Our Knowledge of Channel Island Abalone Resources," nd, 1, Mia Tegner Papers, Box 15, folder Ab/Urchin Literature Review, SIOA.

44 CFGC, meeting minutes, January 17, 1975, 25.

45 Burge, interview, 2011.

46 Burge, interview, 2011.

47 Cicin-Sain et al., "Management Approaches," 77, 79.

48 Cicin-Sain et al., "Management Approaches," 110–111.

CHAPTER 9

1 Charles Lincoln Edwards, "The Abalones of California," *Popular Science Monthly* 82 (June 1913): 549; John G. Carlisle Jr., "The Technique of Inducing Spawning in *Haliotis rufescens* Swainson," *Science* 102 (November 1945): 566–567; Keith W. Cox, "Report of Visit to Abalone Hatcheries in Japan," CDFG, Marine Resources Operations Laboratory, March 1963, 1–5; Earl E. Ebert, "Abalone Aquaculture: A North American Regional Overview," in *Abalone of the World: Biology, Fisheries and Culture*, Proceedings of the First International Symposium on Abalone, ed. S. A. Shepherd, M. J. Tegner, and S. Guzmán del Próo (Blackwell Scientific, 1992), 570–582; A. L. "Scrap" Lundy, *The California Abalone Industry, a Pictorial History* (Best, 1997), 193–195.

2 Jim Lichatowich, *Salmon without Rivers: A History of the Pacific Salmon Crisis* (Island Press, 1999), 148–150. Ironically, fish biologists had already begun to grasp how hatcheries could inadvertently diminish the genetic diversity and thereby the productivity of fishes, which led some to start questioning the long-term efficacy of expensive fish-culture programs, a matter of vigorous discourse that continues to this day.

3 Daniel Morse, interview, February 5, 2014.

4 Morse, interview, 2014; Aileen A. C. Morse, "Algae Make Abalone Settle Down," *Science News* 115, no. 23 (1979): 374; Aileen A. C. Morse, "How Do Planktonic Larvae Know Where to Settle?" *American Scientist* 79, no. 2 (1991): 154–167.

5 Daniel E. Morse et al., "Hydrogen Peroxide Induces Spawning in Mollusks, with Activation of Prostaglandin Endoperoxide Synthetase," *Science* 196 (April 1977): 298–300; Daniel E. Morse et al., "*Gamma*-Aminobutyric Acid, a Neurotransmitter, Induces Planktonic Abalone Larvae to Settle and Begin Metamorphosis," *Science* 204, no. 4391 (1979): 407–410.

6 John Richards, interview, March 14, 2014; John McKenzie, "Studies Underway to Increase Growth and Survival of Abalone: A Three-Year Program," *Outdoor California* 38, no. 3 (1977): 28–29.

7 Laura Rogers-Bennett, "In Memoriam: Mia Jean Tegner, 1947–2001," *CalCOFI Report* 47 (2002): 5; Paul K. Dayton, interview, April 2, 2011.

8 Brock Bernstein and Richard W. Welsford, "An Assessment of Feasibility of Using High-Calcium Quicklime as an Experimental Tool for Research into Kelp Bed/Sea Urchin Ecosystems in Nova Scotia," *Canadian Technical Report of Fisheries and Aquatic Sciences* 968 (1982): 1.

9 Susumu Kato and Stephen C. Schroeter, "Biology of the Red Sea Urchin, *Strongylocentrotus franciscanus*, and Its Fishery in California," *Marine Fisheries Review* 47, no. 3 (1985): 1–2.

10 Dayton, interview, 2011.

11 Mia J. Tegner and Paul K. Dayton, "Sea Urchin Recruitment Patterns and Implications of Commercial Fishing," *Science* 196 (April 1977): 324–326.

12 Wheeler J. North and John S. Pearse, "Sea Urchin Population Explosion in Southern California Coastal Waters," *Science* 167, no. 3915 (1970): 209; Mia J. Tegner and Lisa Levin, "Do Sea Urchins and Abalones Compete in California Kelp Forest Communities?" in *International Echinoderm Conference, Tampa Bay*, Proceedings (1982), ed. J. M. Lawrence, 268; David L. Leighton, "Grazing Activities of Benthic Invertebrates in Southern California Kelp Beds," in special issue, "The Biology of Giant Kelp Beds (*Macrocystis*) in California," ed. W. J. North, *Beihefte zur Nova Hedwigia* 32 (1971): 421–453. Decades earlier, commercial diver Ernie Porter had described this secondary relationship when he voiced concern that sea otters would eliminate abalone nursery habitat by eating large red urchins; CDFG biologist Earl Ebert had also observed this secondary relationship.

13 Tegner and Levin, "Do Sea Urchins," 270; T. A. Dean et al., "Effects of Grazing by Two Species of Sea Urchins (*Strongylocentrotus franciscanus* and *Lytechinus anamesus*) on Recruitment and Survival of Two Species of Kelp (*Macrocystis pyrifera* and *Pterygophora californica*)," *Marine Biology* 78, no. 3 (1984): 301–313.

14 Tegner and Levin, "Do Sea Urchins," 265–270.

15 Mia Tegner, "An Analysis of Our Knowledge of Channel Island Abalone Resources," nd, 1, Mia Tegner Papers, Box 15, folder Ab/Urchin Literature Review, SIOA.

16 Parke H. Young, "Some Effects of Sewer Effluent on Marine Life," *California Fish and Game* 50 (1964): 33–41; Alan J. Mearns et al., "Recovery of Kelp Forest off Palos Verdes," Southern California Coastal Water Research Project, Annual Report (1977), 98–108, ftp.sccwrp.org/pub/download/DOCUMENTS/AnnualReports/1977AnnualReport/ar12.pdf.

17 Mia J. Tegner and Robert A. Butler, "The Survival and Mortality of Seeded and Native Red Abalones, *Haliotis rufescens*, on the Palos Verdes Peninsula," *California Fish and Game* 71, no. 3 (1985): 150–163.

18 Tegner and Butler, "Survival," 153–154, 161.

19 Tegner and Butler, "Survival," 152–153; Mia Tegner, "Abalone Seminar," Tegner Papers, Box 15, folder IOLR, July 1988, SIOA; Eric Hanauer, interview, May 20, 2011; Richard DeFelice, CDFG Cruise Report, 79-x-10, November 26–30, 1979, Experimental Abalone Enhancement Program.

20 Tegner and Butler, "Survival," 155.

21 Tegner and Butler, "Survival," 156.

22 Kristine Barsky (formerly Kristine Henderson), interview, March 4, 2014.

23 Thomas B. McCormick III and Kirk O. Hahn, "Japanese Abalone Culture Practices and Estimated Costs of Juvenile Production in the USA," *Journal of World Mariculture Society* 14 (1983): 149. Konstantin Karpov, interview, April 2, 2010. McCormick estimated the cost for a 2 cm seed abalone to be 44 cents in 1983; Karpov estimated the cost at greater than $400 per abalone. Tegner was planting out 2- to 3-inch abalone, so costs would have been higher.

24 Mia Tegner, "Santa Barbara Abalone Enhancement Talk," August 2, 1981, in Tegner Papers, SIOA; Eric Hanauer, interview, 2011.

25 Tegner, "Santa Barbara Abalone"; Paul Dayton, correspondence, September 3, 2019. Tegner additionally explained that in Japan, abalone divers used less intensive dive technology and were organized into locally based co-ops that closely controlled the seasons, techniques, and size of abalone harvests. She also noted the lack of scientific documentation about the success of Japan's abalone hatcheries.

26 Mia J. Tegner et al., "Experimental Abalone Enhancement Program," in California Sea Grant College Program 1978–1980, *Biennial Report*, Institute of Marine Resources, UC La Jolla (1981), 114–116; R. J. Schmitt and J. H. Connell, "Field Evaluation of an Abalone Enhancement Program," in California Sea Grant College Program 1980–1982, *Biennial Report*, Institute of Marine Resources, UC La Jolla (1982), 172–176; McKenzie, "Studies Underway," 28–29; Ronald S. Burton and Mia J. Tegner, "Enhancement of Red Abalone *Haliotis rufescens* Stocks at San Miguel Island: Reassessing a Success Story," *Marine Ecology Progress Series* 202 (August 2000): 303–308; Tegner, "Abalone Seminar," SIOA; Richard DeFelice, CDFG Cruise Report, 79-x-7, March 26–30, 1979, Experimental Abalone Enhancement Program; Richard DeFelice, CDFG Cruise Report, 79-x-9, September 17–21, 1979, Experimental Abalone Enhancement Program; Richard DeFelice, CDFG Cruise Report, 79-x-10, November 26–30, 1979, Experimental Abalone Enhancement Program; Robert Butler, CDFG Cruise Report, 80-x-15, May 5–9, 1980, Experimental Abalone Enhancement Program; Robert Butler and Mia Tegner, CDFG Cruise Report, 80-x-17, September 2–4, 1980, Experimental Abalone Enhancement Program; David O. Parker, CDFG Cruise Report, 83-x-4, May 18–19, 1983, Energy Resources Fund Abalone Enhancement Program; Kristine C. Henderson, CDFG Cruise Report, 83-x-5, June 14–15, 1983, Energy Resources Fund Abalone Enhancement Program.

27 Tegner, "Abalone Seminar."

28 Mia J. Tegner and Robert A. Butler, "Drift-Tube Study of the Dispersal Potential of Green Abalone (*Haliotis fulgens*) Larvae in the Southern California Bight: Implications for Recovery of Depleted Populations," *Marine Ecology Progress Series* 26 (1985): 73–84.

29 Tegner and Butler, "Drift-Tube Study," 75; David L. Leighton et al., "Acceleration of Development and Growth in Young Green Abalone (*Haliotis fulgens*) Using Warmed Effluent Water," *Journal of World Mariculture Society* 12 (1981): 170–180.

30 "Drifting Bottles Help in Study of Ways to Increase Abalone," *Los Angeles Times*, November 4, 1981.

31 Tegner and Butler, "Drift-Tube Study," 80–83; Tegner, "Abalone Seminar."

32 Mia J. Tegner, "Brood-Stock Transplants as an Approach to Abalone Stock Enhancement," in *Abalone of the World: Biology, Fisheries and Culture, Proceedings of the First International Symposium on Abalone*, ed. S. A. Shepherd, Mia J. Tegner, and S. A. Guzmán del Próo (Fishing News Books, 1992), 461.

33 Tegner, "Brood-Stock Transplants," 461–474; Gordon Grant, "The Great Abalone Roundup: Dwindling Delicacy Is Transplanted," *Los Angeles Times*, August 9, 1982.

34 Tegner, "Brood-Stock Transplants," 467; Tom Knudson and Nancy Vogel, "Secret Slaughter by Poachers Takes Huge Toll on Wildlife," *Sacramento Bee*, December 24, 1996.

35 Tegner, "Brood-Stock Transplants," 467.

36 Knudson and Vogel, A1; Tegner, "Brood-Stock Transplants," 472.

37 Knudson and Vogel, "Secret Slaughter."

38 Knudson and Vogel, "Secret Slaughter" (quote); Mia J. Tegner, "Abalone (*Haliotis* spp.) Enhancement in California: What We've Learned and Where We Go from Here," in *Workshop on Rebuilding Abalone Stocks in British Columbia*, Canadian Special Publication of Fisheries and Aquatic Sciences, ed. Alan Campbell (NRC Research Press 2000), 67; Kristine C. Henderson et al., "The Survival and Growth of Transplanted Adult Pink Abalone, *Haliotis corrugata*, at Santa Catalina Island," *California Fish and Game* 74, no. 2 (1988): 82–86.

39 Burton and Tegner, "Enhancement," 306; Tegner, "Abalone (*Haliotis* spp.) Enhancement," 69. Tegner would later recommend abalone transplants for Marine Protected Areas, where presumably they could be safeguarded from poachers.

40 Tegner, "Abalone Seminar."

CHAPTER 10

1   Mia J. Tegner et al., "Population Biology of Red Abalones, *Haliotis rufescens*, in Southern California and Management of the Red and Pink, *H. corrugata*, Abalone Fisheries," *Fishery Bulletin* 87, no. 2 (1989): 330.

2   Peter L. Haaker et al., "Red Abalone Size Data from Johnsons Lee, Santa Rosa Island, Collected from 1978 to 1984," CDFG, Marine Resources Technical Report 53 (1986):3. Humboldt State University professor John DeMartini, Burge's grad school mentor, already had similar research under way in northern California.

3   Tegner et al., "Population Biology," 314; Peter L. Haaker et al., "Growth of Red Abalone, *Haliotis Rufescens* (Swainson), at Johnsons Lee, Santa Rosa Island, California," *Journal of Shellfish Research* 7, no. 3 (1998): 752.

4   Michael Huston et al., "New Computer Models Unify Ecological Theory," *BioScience* 38, no. 10 (1988): 682.

5   Tegner et al., "Population Biology," 314–315, 322–323.

6   Tegner et al., "Population Biology," 331–332.

7   Haaker et al., "Growth of Red Abalone," 751.

8   Haaker et al., "Growth of Red Abalone," 752; Paul Dayton and Mia Tegner, "Catastrophic Storms, El Niño, and Patch Stability in a Southern California Kelp Community, *Science* 224 (1984): 283–285; Larry Basch, interview, October 20, 2018.

9   S. George Philander, *El Niño, La Niña, and the Southern Oscillation* (Academic Press, 1990), 2–9; William J. Emery and Kevin Hamilton, "Atmospheric Forcing of Interannual Variability in the Northeast Pacific Ocean: Connections with El Niño," *Journal of Geophysical Research* 90 (1985): 857–868.

10 Tegner et al., "Population Biology," 336.

11 Mia J. Tegner and Paul Dayton, "El Niño Effects on Southern California Kelp Forest Communities," *Advances in Ecological Research* 17 (1987): 243–275; Peter Haaker, interview, February 18, 2010.

12 Tegner et al., "Population Biology," 336.

13 Tegner et al., "Population Biology," 330.

14 Mia J. Tegner et al., "Climate Variability, Kelp Forests, and the Southern California Red Abalone Fishery," *Journal of Shellfish Research* 20, no. 2 (2001): 760–761.

15 Peter Haaker, interview, April 30, 2010.

16 Konstantin Karpov et al., "The Red Abalone, *Haliotis Rufescens*, in California: Importance of Depth Refuge to Abalone Management," *Journal of Shellfish Research* 17, no. 3 (1998): 866.

17 Peter Haaker and Konstantin Karpov, interview via Skype, October 9, 2011. In 1995, Haaker and Karpov did twenty random transects at Johnsons Lee and found only six abalone; at a nearby NPS monitoring site, red abalone numbers dropped to zero.

18 Ian Taniguchi, CDFG Cruise Report, 01-M-3, April 23 to May 2001, Nearshore Invertebrates.

19 Gary Davis, interview, May 3, 2010; Haaker, interview, April 30, 2010; *United States v. California*, 436 U.S. 32 (1978); "Enabling Legislation for Channel Islands National Park, Public Law 96-199, 96th Congress, March 5, 1980," in *Foundation Document: Channel Islands National Park, California*, NPS (2017), Appendix A; Gary E. Davis, "National Park Stewardship and Vital Signs Monitoring: A Case Study from Channel Islands National Park, California," *Aquatic Conservation: Marine and Freshwater Ecosystems* 15 (2005): 71–89.

20 Dan Richards, interview, March 18, 2010; Haaker, interview, April 30, 2010; Basch, interview, 2018; field biologists from the Channel Islands Research Program also helped to establish the CINP program's monitoring sites and to collect initial data.

21 Davis, interview, 2010; Dan Richards, interview, April 17–22, 2010; Haaker, interview, April 30, 2010.

22 Gary E. Davis, "Mysterious Demise of Southern California Black Abalone, *Haliotis cracherodii* Leach, 1814," *Journal of Shellfish Research* 12, no. 2 (1993): 183–184.

23 Brian Tissot, "Recruitment, Growth, and Survivorship of Black Abalone on Santa Cruz Island Following Mass Mortality," *Bulletin of the Southern California Academy of Sciences* 94, no. 3 (1995): 179–189; Daniel V. Richards, "Rocky Intertidal Ecological Monitoring in Channel Islands National Park, California, 1986–87," *CINP Natural Science Reports*, CHIS-88-001, 9.

24 Pete Huisveld, *Abalone from Sea to Saucepan* (Diving Dirtball, 1975), 20–22; Albert Yu-Min Lin et al., "Underwater Adhesion of Abalone: The Role of Van der Waals and Capillary Forces," *Acta Materialia* 57 (2009): 4178. Four hundred pounds of force comes from Huisveld; Lin drew a similar conclusion: that a 6-inch red abalone can withstand a 115kPa pull-off force, equivalent to 16.6 pounds of force per square inch, or 469 pounds of force.

25 Peter L. Haaker et al., "Mass Mortality and Withering Syndrome in Black Abalone, *Haliotis cracherodii*," in *Abalone of the World: Biology, Fisheries and Culture, Proceedings of the First International Symposium on Abalone*, ed. Scoresby A. Shepherd, Mia J. Tegner, and S. A. Guzmán del Próo (Fishing News Books, 1992), 215–224.

26 Drew Harvell, *Ocean Outbreak: Confronting the Rising Tide of Marine Disease* (University of California Press, 2019), 67–69.

27 J. B. Blecha et al., "Aspects of the Biology of the Black Abalone (*Haliotis cracherodii*) near Diablo Canyon, Central California," in *Abalone of the World*, 128–136.

28 Davis, "Mysterious Demise," 183–184.

CHAPTER 11

1 Jeremy D. Prince and Scoresby A. Shepherd, "Australian Abalone Fisheries and Their Management," in *Abalone of the World: Biology, Fisheries, Culture, Proceedings of the First International Symposium on Abalone*, ed. S. A. Shepherd, Mia J. Tegner, S. A. Guzman Del Próo (Fishing News Books, 1992), 407–426.

2 Scoresby Shepherd, correspondence with author, September 14, 2011.

3 Shepherd et al., *Abalone of the World*.

4 Scoresby A. Shepherd, "Studies on Southern Australian Abalone (Genus *Haliotis*) VII: Aggregative Behaviour of *H. laevigata* in Relation to Spawning," *Marine Biology* 90, no. 2 (1986): 231–236; S. A. Shepherd, "Movement of the Southern Australian Abalone *Haliotis laevigata* in Relation to Crevice Abundance," *Australian Journal of Ecology* 11, no. 3 (1986): 295–302.

5 Don R. Levitan and Tamara M. McGovern, "The Allee Effect in the Sea," in *Marine Conservation Biology: The Science of Maintaining the Sea's Biodiversity*, ed. Elliott Norse and Larry B. Crowder (Island Press, 2005), 47.

6 Jeremy D. Prince, "Using a Spatial Model to Explore the Dynamics of an Exploited Stock of the Abalone *Haliotis rubra*," in *Abalone of the World*, 305–317.

7 Gary Davis, interview, May 3, 2010.

8 Peter Haaker, interview, April 30, 2010.

9 CFGC, meeting minutes, August 1–2, 1991, CSA.

10 CDFG, Bill Analysis AB 3705, April 17, 1990, CSA, Bill File AB 3705, Senate Natural Resources and Wildlife Committee, LP383:94; Mike Kitahara (California Abalone Association) to Rep. Dan Hauser, letter, April 23, 1990, CSA, Bill File AB3705, Rep. Dan Hauser, LP367:219-220; CDFG, Marine Region, *ARMP*, Appendix A, "Historical Summary of Laws and Regulations Governing the Abalone Fishery in California," A7–A17.

11 Haaker, interview, April 30, 2010; CDFG, "Draft Environmental Document, Black Abalone Fishery Closure" (March 1993), 1.3, 2.2, 7.1.

12 CDFG, "Final Environmental Document, Black Abalone Fishery Closure" (1993), 4.3.

13 CDFG, "Final Environmental Document," 1.2.

14 John Colgate to CFGC, letter, April 23, 1993; all commercial divers' letters included in "Final Environmental Document," 8.3.

15 Jim Marshall to CFGC, letter, April 12, 1993, in "Final Environmental Document," 8.3.

16 Jim Marshall to CFGC, letter, 1993.

17 Steve Rebuck (CAA) to Robert Treanor (CFGC), letter, May 3, 1993, in "Final Environmental Document," 8.3.

18 CFGC, meeting minutes, June 17, 1993, 26–27, CSA.

19 CFGC, meeting minutes, June 17, 1993, 26–27.

20 CFGC, meeting minutes, June 17, 1993, 27.

21 CFGC, meeting minutes, June 17, 1993, 26.

22 CFGC, meeting minutes, June 17, 1993, 28.

23 Haaker, interview, April 30, 2010.

24 CDFG, "Pink, Green, and White Abalone Fishery Closure, Draft Environmental Document" (1995), 3.112-3.113.

25 Dan Richards, interview, March 18, 2010.

26 Haaker, interview, April 30, 2010.

27 Paul Bartsch, "The West American Haliotis," *Proceedings of the U.S. National Museum* 89, no. 3094 (1940): 49–58; Kevin D. Lafferty, "Restoration of the White Abalone in Southern California: Population Assessment, Brood Stock Collection, and Development of Husbandry Technology, Final Report," USGS, Western Ecological Research Center, unpublished grant report (2001), 10; Buzz Owen (retired commercial diver and abalone enthusiast), interview by David Kushner (NPS biologist), May 1, 2001; David Kushner (NPS biologist), interview and correspondence, March 18, 2010.

28 Davis, interview, February 19, 2010.

29 Davis, interview, February 19, 2010.

30 Gary E. Davis et al., "Status and Trends of White Abalone at the California Channel Islands," *Transactions of the American Fisheries Society* (1996): 42–48.

31 Davis, interview, February 19, 2010; Glen S. Jamieson, "Marine Invertebrate Conservation: Evaluation of Fisheries Over-Exploitation Concerns," *American Zoologist* 33, no. 6 (1993): 551.

32 Peter L. Haaker, "White Abalone—Off the Deep End . . . Forever?" *Outdoor California* 59, no. 1 (1998): 18–20; Konstantin Karpov, "White Abalone—An Extinct Possibility, *Kelp Forest Newsletter* (May 1998): 4–5.

33 Davis, interview, May 3, 2010.

34 Haaker, "White Abalone," 19.

35 Haaker, "White Abalone," 19.

36 Haaker, "White Abalone," 19.

37 NPS, "The Race to Save the White Abalone," film, nd. This film can be seen at the CINP website, http://www.nps.gov/chis/photosmultimedia/race-to-save-the-white-abalone.htm.

38 Davis, interview, May 3, 2010.

39 Davis, interview, May 3, 2010.

40 Haaker, "White Abalone," 20.

41 Gary E. Davis et al., "The Perilous Condition of White Abalone (*Haliotis sorenseni*, Bartsch, 1940)," *Journal of Shellfish Research* 17, no. 3 (1998): 871–876.

42 NPS, "The Race to Save the White Abalone," film.

43 Davis et al., "Status and Trends," 1996, 46.

44 Davis, interview, May 3, 2010; Mia J. Tegner et al., "Near Extinction of an Exploited Marine Invertebrate," *Trends in Ecology and Evolution* 11, no. 7 (1996): 27.

45 Davis et al., "Status and Trends," 1996, 45–46.

46 Davis et al., "Status and Trends," 1996, 43.

47 Davis et al., "Status and Trends," 1996, 43.

48 Davis et al., Status and Trends," 1996, 43. Buzz Owens would later report that that white abalone had been harvested earlier but counted as pinks; Owens, an avid shell collector, also affirmed that juvenile white abalone were already very difficult to find by the mid-1960s; Kushner interview.

49 Yvette La Pierre, "Shell Game," *National Parks Magazine* (Sept/Oct 1998), 39.

50 Alistair J. Hobday et al., "Over-Exploitation of a Broadcast Spawning Marine Invertebrate: Decline of the White Abalone," *Reviews in Fish Biology and Fisheries* 10 (2001): 506; Kevin Lafferty, interview, January 30, 2014; Tegner et al., "Near Extinction," 27.

51 Davis et al., "Perilous Condition," 872–873.

52 Davis et al., "Status and Trends," 1996, 42–48.

53 Davis et al., "Perilous Condition," 875.
54 CDFG, "Pink, Green, and White Fishery" (1995), 1–5.
55 CFGC, meeting minutes, November 3, 1995, 50; CFGC, meeting minutes, December 8, 1995, 41–43, both in CSA.
56 CFGC, meeting minutes, December 8, 1995, 44.

CHAPTER 12

1 Tom Stienstra, "Turnbull's Cries for a Crackdown on Abalone Diving Heard," *San Francisco Chronicle*, May 12, 1997; "Abalone Records and Stats," post by Matt Mattison, January 21, 2011, Spearboard Spearfishing Community, http://spearboard.com/showthread.php?t=118423; Kurt Bickel, "A History of Abalone Harvest Regulations in California," *SCAN Newsletter* (October 2003), 5; Steve Campi (sport diver and former Cen-Cal president), interview, February 15, 2014.
2 Stienstra, "Turnbull's Cries Heard."
3 Stienstra, "Turnbull's Cries Heard."
4 CFGC, meeting minutes, December 5–6, 1996, 16; CFGC meeting minutes, February 6–7, 1997, 10, both in CSA.
5 Rocky Daniels (sport diver and former SCAN president), interview, March 12, 2010.
6 Rocky Daniels, correspondence with author, July 28, 2019.
7 Konstantin Karpov et al., "The Red Abalone, *Haliotis rufescens*, in California: Importance of Depth Refuge to Abalone Management," *Journal of Shellfish Research* 17, no. 3 (1998), 863–870.
8 Jerald S. Ault and John D. Martini, "Movement and Dispersion of Red Abalone, *Haliotis rufescens*," *California Fish and Game* 73, no. 4 (1987): 196–213.
9 Tom Stienstra, "Abalone Poaching Leaves Scar," *San Francisco Examiner*, December 8, 1996; Alex Barnum, "Panel Delays Action on Abalone Harvest: Biologists Say Commercial Ban Needed," *San Francisco Chronicle*, March 7, 1997.
10 Bickel, "History," 5; Daniels, interview, 2010.
11 Tom Stienstra, "Abalone Poachers Caught, but the Damage Lingers On," *San Francisco Examiner*, December 8, 1996.
12 Konstantin Karpov, "Poached Abalone: Recipe for Disaster," *Kelp Forest Newsletter* (January 1995): 3; Louis Martin, "Frustration, Anger over Abalone Poaching Plea Bargain," *Coast News*, December 6, 1995.
13 Brook Halsey, interview, September 25, 2019; Martin, "Frustration"; Louis Martin, "Lenient Plea Bargain Withdrawn by Judge, Abalone Poaching Case Set for Trial," *Coast News*, February 6, 1996.
14 Daniels, interview, 2010; Tom Stienstra, "Wardens Poach a Big Subject," *San Francisco Chronicle and Examiner*, June 8, 1997; Halsey, interview, 2019.
15 Louis Martin, "Abalone Poacher 'Mastermind' Found Guilty on All Counts," *Coast News*, May 28, 1996; Louis Martin, "Pressure Grows for Tougher Sentences for Abalone Poachers," *Coast News*, January 26, 1996; Trip Gabriel, "Stewart's Point Journal; Valuable California Abalone Draws Sportsmen and Poachers," *New York Times*, June 22, 1996.
16 Konstantin A. Karpov et al., "Interactions among Red Abalones and Sea Urchins in Fished and Reserve Sites of Northern California: Implications of Competition to Management," *Journal of Shellfish Research* 20, no. 2 (2001): 743–753. Northern California was opened to commercial urchin fishing in 1985.
17 Glen Martin, "Stopping the Abalone Shell Game," *San Francisco Chronicle*, October 15, 1990.
18 Karpov, "Poached Abalone," 3.
19 Karpov, "Poached Abalone," 3; Rocky Daniels and Rand Floren, "Poaching Pressures on Northern California's Abalone Fishery," *Journal of Shellfish Research* 17, no. 3 (1998): 859–862; Will Harper, "A Dying Breed," *East Bay Express*, January 23, 2002; Bickel, "History."
20 Karpov, "Poached Abalone," 3; Daniels and Floren, "Poaching Pressures," 861.
21 Dan Hauser, interview, September 23, 2019; Bill Analysis AB3705, CSA, Bill File AB3705, Hauser; CFGC, meeting minutes, March 6, 1991, 38–39, CSA.

22 Konstantin Karpov, "Marine Reserve Dive Survey Results," *Kelp Forest Newsletter* (January 1994): 3; Konstantin Karpov, interview, October 10, 2011.

23 Bickel, "History," 5; "Northern California Sportfishing," *Kelp Forest Newsletter* (October 1992): 1. In 1992, Karpov started CDFG's *Kelp Forest, Northern California Marine Sport Fishing and Diving Newsletter*, to inform the public about the status of marine sportfishery resources.

24 Konstantin A. Karpov et al., "Serial Depletion and the Collapse of the California Abalone (*Haliotis* spp.) Fishery," in *Workshop on Rebuilding Abalone Stocks in British Columbia*, ed. Alan Campbell, Canadian Special Publication in Fisheries and Aquatic Science, 130 (NRC Research Press, 2000), 11–24; Konstantin A. Karpov, "Increased Commercial Harvest of Red Abalone, *Haliotis rufescens* (Swainson) from the San Francisco Area in Relation to the South-Central California Fishery: Is Poaching a Factor? With Proposed Regulation Changes and Management Strategies," draft (May 10, 1989) in CDFG, Marine Resources Division, *Final Supplemental Environmental Document, Abalone Ocean Sportfishing* (1991), Appendix-150.

25 Steve Benavides, interview, February 17, 2011.

26 Marla Cone, "Prop. 132: Allen Nets One Victory for Environment," *Los Angeles Times*, November 8, 1990.

27 Campi, interview, 2014.

28 Tom Stienstra, "Nothing Phony about Efforts to Save Abalone," *San Francisco Examiner*, December 1, 1996.

29 Tom Stienstra, "Abalone Battle Put on Shelf until February," *San Francisco Chronicle*, December 10, 1996; Steve Benavides, interview, July 19, 2019; Rocky Daniels, interview, July 15, 2019.

30 Paul J. Turnbull (AMRC) to Robert Treanor (director, CFGC), letter, October 18, 1996, CSA, Bill File SB 463, Sen. Mike Thompson, 1997, LP383.288.

31 Turnbull to Treanor, letter, October 18, 1996.

32 CFGC, meeting minutes, December 5–6, 1996, 16; "1996/97 Abalone Harvesting Closure," from Rocky Pages, website, http://abdiving.com/iclosure.htm, accessed June 15, 2011; Rocky Daniels, interview, 2019.

33 Stienstra, "Nothing Phony." Quotes are from testimony provided to reporter Tom Stienstra ahead of the meeting.

34 Stienstra, "Nothing Phony."

35 CFGC, meeting minutes, December 5–6, 1997; Al Petrovich, interview, June 21, 2011; Rocky Daniels, correspondence with author, July 28, 2019; Benavides, interview, 2019.

36 Tom Knudson and Nancy Vogel, "Secret Slaughter by Poachers Takes Huge Toll on Wildlife," *Sacramento Bee*, December 24, 1996, 1.

37 Tom Knudson and Nancy Vogel, "Californians Are Squandering Their Coastal Heritage," *Sacramento Bee*, December 22, 1996, 1.

38 Benavides, interview, 2011. In 1991–1992, the CAA had continued its press to open an experimental fishery on the North Coast, kindling continued concern from sport divers and local communities. CFGC, meeting minutes, March 6, 1992.

39 Daniels, interview, 2010; Benavides, interview, 2011; Campi, interview, 2014.

40 Benavides, interview, 2011.

41 CFGC, meeting minutes, February 6–7, 1997, 10–15.

42 CFGC, meeting minutes, February 6–7, 1997, 10–15.

43 CFGC, meeting minutes, February 6–7, 1997, 10–15.

44 Alex Barnum, "Abalone Decision Postponed: Biologists, Divers Say Commercial Ban Needed," *San Francisco Chronicle*, March 8, 1997; Konstantin A. Karpov and Peter Haaker, "A Decline without Refuge? The Red Abalone and Status of Populations in California," *Kelp Forest Newsletter* (April 1997): 8.

45 CFGC, meeting minutes, February 6–7, 1997, 11; March 6–7, 1997, 9, CSA; *Channel Islands Council of Divers*, newsletter, nd, CSA, Bill File SB 463, Chapter 787, Chris Chauncey, Abalone (October 8, 1997), 1998-11-33.

46 CFGC, meeting minutes, March 6–7, 1997, 9.

47 Newspaper clipping, nt, nd, CSA, Bill File SB 463, Senate Committee on Natural Resources and Wildlife, LP383.288.

48 CFGC, meeting minutes, March 6–7, 1997, 8–9.

49 Alex Barnum, "Abalone Decision."

50 "California Abalone Season in Doubt—State Takes First Step toward Diving Ban to Preserve Prized Mollusk," *Sacramento Bee*, March 7, 1997.

51 CFGC, meeting minutes, March 6–7, 1997, 10–13; CFGC, meeting minutes, May 6–7, 1997, 3, CSA.

52 "Bill Would Ban Abalone Kills," *Sacramento Bee*, April 10, 1997; Benavides, interview, 2011; Daniels, interview, 2010; "Important Facts about California's Red Abalone Resource 1996," CSA, Bill File SB 463, Chapter 787 (October 8, 1997); Mike Henderson to Ruth Coleman and Chris Chauncy (Sen. Thompson staffers), letter, April 8, 1997, CSA, Bill File SB 463; Mark J. Spalding, "California and the World Ocean '97: Taking a Look at California's Ocean Resources: An Agenda for the Future," *Journal of Environment and Development* 6, no. 4 (1997): 453–455; "Swept Up in a Wave of Reform," *Sacramento Bee*, March 14, 1997.

53 Mia Tegner to Senator Mike Thompson, letter, April 10, 1997, CSA, Bill File SB 463, Thompson.

54 Philip Stark, "A Case Study in Natural Resource Legislation," Chapter 32 in *SticiGui* (online statistics textbook), http://www.stat.berkeley.edu/~stark/SticiGui/Text/abalone.htm, accessed February 17, 2010; Graphs from Stark's presentation, CSA, Bill File SB 463, Senate Committee on Natural Resources and Wildlife.

55 Daniels, interview, 2010; Campi, interview, 2014.

56 Letter of support from dive club leaders (Steve Campi, Cen-Cal; Karen Ross, San Diego Council of Divers; Amy Anders, GLACD; Keni Cyr-Rumble, Sport Chalet Dive Clubs; Frank King, San Francisco Reef Divers; David Sommers, Redwood Empire Divers; George Tutwiler, JAWS Dive Club; George Lawry, SCAN) to Sen. Mike Thompson, April 16, 1997; Gene Kramer to Sen. Mike Thompson, June 24, 1997; Stephen G. Benavides to Sen. Mike Thompson, letter, April 29, 1997, all in CSA, Bill File SB 463, Thompson.

57 Daniels, interview, 2010.

58 CFGC, meeting minutes, May 6–7, 1997, 4–5.

59 CFGC, meeting minutes, May 6–7, 1997, 5.

60 "Abalone Divers Face Sharks, Tides, and Now Officials," *New York Times*, May 9, 1997.

61 CFGC, meeting minutes, May 6–7, 1997, 3; Fred Wendell, "California Department of Fish and Game's Response to the Sabellid Worm Infestation," in *Biological Invasions in Aquatic Ecosystems: Impacts on Restoration and Potential for Control, Proceedings of a Workshop*, April 25, 1998, Sacramento, ed. A. N. Cohen and S. K. Webb (San Francisco Estuary Institute, 2002), 8.

62 "Red Abalone Harvest Temporarily Banned," *San Jose Mercury News*, May 7, 1997.

63 CFGC, meeting minutes, May 6–7, 1997, 5.

64 CFGC, meeting minutes, May 6–7, 1997, 6.

65 CFGC, meeting minutes, May 6–7, 1997, 6; Stark, "Case Study."

66 CFGC, meeting minutes, May 6–7, 1997, 5–6.

67 CFGC, meeting minutes, May 6–7, 1997, 7; Steve Rebuck, interview, March 26, 2014.

68 CFGC, meeting minutes, May 6–7, 1997, 7.

69 Peter Haaker, interview, February 18, 2010.

70 Arthur F. McEvoy, *The Fisherman's Problem: Ecology and Law in the California Fisheries, 1850–1980* (Cambridge University Press, 1986), 153–155, 215–217; David Dobbs, *The Great Gulf: Fishermen, Scientists, and the Struggle to Revive the World's Greatest Fishery* (Island Press, 2000), 58–82.

71 "Red Abalone Harvest," *San Jose Mercury News*.

72 "Red Abalone Harvest," *San Jose Mercury News*.

73 "Abalone Divers," *New York Times*.

74 Benavides, interview, 2011.

75 Stienstra, "Wardens Poach"; Bony Saludes, "Ton of Sonoma County Abalone Seized in Poaching Sting," *Santa Rosa Press Democrat*, May 30, 1997.

76 Gordon Gregg to Sen. Mike Thompson, letter, May 5, 1997, CSA, Bill File SB 463, Thompson.

77 John Colgate and Steve Rebuck, "Draft: Abalone Reserves and Closures, Combination of New and Colgate Ideas," April 25, 1997; "Draft Red Abalone Restoration Plan," June 23, 1997, CSA, Bill File SB 463, Assembly Water, Parks, and Wildlife Committee, LP361:290.

78 California Legislative Information, SB 463, http://info.sen.ca.gov/pub/97-98/bill/sen/sb_0451-0500/sb_463_bill_19970911_enrolled.html, accessed March 28, 2010.

79 "Abalone Moratorium Passes," *Sacramento Bee*, September 12, 1997; "Wilson Signs Bills to Protect Coast, Abalone Industry Put on Hold," *Sacramento Bee*, October 9, 1997.

CHAPTER 13

1 Gary E. Davis et al., "The Perilous Condition of White Abalone *Haliotis Sorenseni*, Bartsch, 1940," *Journal of Shellfish Research* 17, no. 3 (1998): 871–875.

2 Davis et al., "Perilous Condition," 874–875.

3 Gary Davis, interview, May 3, 2010; NMFS, "Endangered and Threatened Wildlife and Plants; 90-Day Finding for a Petition to List White Abalone (*Haliotis sorenseni*) as Endangered," *Federal Register* 64, no. 185 (1999): 51725–51727.

4 Davis, interview, 2010; John Butler, interview, February 1, 2014.

5 Alistair J. Hobday and Mia J. Tegner, "Status Review of White Abalone (*Haliotis sorenseni*) throughout Its Range in California and Mexico," NOAA Technical Memorandum (May 2000), 1; Kevin D. Lafferty, "Restoration of the White Abalone in Southern California: Population Assessment, Brood Stock Collection, and Development of Husbandry Technology, Final Report," USGS, Western Ecological Research Center, grant report (December 5, 2001), 4, 8–9; Kevin D. Lafferty et al., "Habitat of Endangered White Abalone, *Haliotis sorenseni*," *Biological Conservation* 116 (2004): 191–194; Butler, interview, 2014.

6 Ian Taniguchi and Peter L. Haaker, CDFG Cruise Report 99-M-10, November 18, 1999, Nearshore Investigations.

7 Ian Taniguchi, CDFG Cruise Report 00-M-3, March 27–29, 2000, Nearshore Investigations.

8 Ian Taniguchi, CDFG Cruise Report 00-M-9b, November 5–9, 2000, Nearshore Investigations; Peter Haaker and Ian Taniguchi, "Fisheries Review: White Abalone," *CalCOFI Reports* 42 (2000): 16–17.

9 Laura Rogers-Bennett, interview, March 17, 2011; Ian Taniguchi, correspondence, September 25, 2019.

10 Hobday and Tegner, "Status Review," 1.

11 Eric Hanauer, "The Perils of Diving for Even the Experienced," Letters to the Editor, *San Diego Union-Tribune*, January 10, 2001.

12 Lisa Petrillo, "Ocean Scientist Mia Tegner Dies while Diving off Mission Beach," *San Diego Union-Tribune*, January 8, 2001.

13 Robert Monroe, "Keepers of the Forest," *Scripps Institution of Oceanography Explorations*, 12, no. 4 (2006): 11.

14 "Obituary Notice Scripps Research Marine Biologist Mia Tegner," Scripps News (online), January 8, 2001; Jack Williams, "Mia Jean Tegner, 53: Marine Biologist," *San Diego Union-Tribune*, January 10, 2001.

15 NMFS, "Endangered and Threatened Species; Endangered Status for White Abalone," *Federal Register*, May 29, 2001, 29046–29055.

16 Davis, interview, 2010.

17 David Leighton, "Laboratory Observations on the Early Growth of the Abalone, *Haliotis sorenseni*, and the Effect of Temperature on Larval Development and Settling Success," *Fishery Bulletin* 70, no. 2 (1972): 373–381.

18 Daniel Morse et al., "Hydrogen Peroxide Induces Spawning in Mollusks, with Activation of Prostaglandin Endoperoxide Synthetase," *Science*, new series 196 (April 1977): 298–300.

19 Daniel Morse, interview, February 5, 2014.

20 Haaker and Taniguchi, "Fisheries Review," 17.

21 USGS, "Team Spawns Rare White Abalone: It's a Girl! It's a Boy! It's 6 Million Baby Mollusks!" press release, April 30, 2001.
22 USGS, "Team Spawns."
23 "Marine Institute Hopes to Bring Back Abalone," *Daily News* (Los Angeles), September 9, 2001.
24 Pete Haaker, interview, April 30, 2010; Melissa Neuman, interview, September 17, 2019.
25 Leighton, "Laboratory Observations," 373; Ronald S. Burton et al., "Restoration of Endangered White Abalone, *Haliotis sorenseni*: Resource Assessment, Genetics, Disease and Culture of Captive Abalone," California Sea Grant College Program, Research Completion Reports (2007), 4–5; Jim Moore, interview, February 7, 2014; Davis, interview, 2010.
26 "White Abalone Recovery Plan (*Haliotis sorenseni*)," NMFS, Office of Protected Resources, Long Beach, CA (October 2008), 1–18; Moore, interview, 2014.
27 "White Abalone Recovery," 1–18; Carolyn S. Friedman et al., "Oxytetracycline as a Tool to Manage and Prevent Losses of Endangered White Abalone, *Haliotis sorenseni*, Caused by Withering Syndrome," *Journal of Shellfish Research* 26, no. 3 (2007): 877–885; Moore, interview, 2014.
28 Peter T. Raimondi et al., "Continued Declines of Black Abalone along the Coast of California: Are Mass Mortalities Related to El Niño Events?" *Marine Ecological Progress Series* 242 (2002): 146; James D. Moore et al., "Withering Syndrome in Farmed Red Abalone *Haliotis rufescens*: Thermal Induction and Association with a Gastrointestinal Rickettsiales-Like Prokaryote," *Journal of Aquatic Animal Health* 12, no. 1 (2000): 26.
29 Carolyn S. Friedman et al., "'*Candidatus xenohaliotis californiensis*' a Newly Described Pathogen of Abalone, *Haliotis* spp. along the West Coast of North America," *International Journal of Systematic and Evolutionary Microbiology* 50 (2000): 847–855.
30 Jim Moore, interview, March 17, 2011.
31 Kevin Lafferty, interview, January 30, 2014; Armand M. Kuris and Carolynn S. Culver, "A Introduced Sabellid Polychaete Pest Infesting Cultured Abalones and Its Potential to Spread to Other California Gastropods," *Invertebrate Biology* 118, no. 4 (1999): 391–403.
32 Fred Wendell, "Withering Syndrome Found in Wild Abalone in the Crescent City Area," *California Marine Currents* 1, no. 3 (1999): 1; Carolyn S. Friedman and Carl A. Finley, "Anthropogenic Introduction of the Etiological Agent of Withering Syndrome into Northern California Abalone Populations via Conservation Efforts," *Canadian Journal of Fisheries and Aquatic Science* 60 (2003): 1424–1431; Moore, interview, 2014.
33 James D. Moore et al., "Withering Syndrome and Restoration of Southern California Abalone Populations," *CalCOFI Reports* 43 (2002): 112–117.
34 John Butler et al., "Status Review Report for Black Abalone (*Haliotis cracherodii* Leach, 1814)" (NMFS, 2009), viii.
35 Pete Raimondi, interview, February 21, 2013.
36 C. Melissa Miner et al., "Recruitment Failure and Shifts in Community Structure Following Mass Mortality Limit Recovery of Black Abalone," *Marine Ecology Progress Series* 327 (2006): 107–117; Raimondi et al., "Continued Declines," 143–152.
37 Raimondi, interview, 2013.
38 D. E. Hamm and R. S. Burton, "Population Genetics of Black Abalone, *Haliotis cracherodii*, along the Central California Coast," *Journal of Experimental Marine Biology and Ecology* 254 (2000): 235–247; K. M. Gruenthal and R. S. Burton, "Genetic Structure of Natural Populations of the California Black Abalone (*Haliotis cracherodii* Leach, 1814), a Candidate for Endangered Species Status," *Journal of Marine Biology and Ecology* 355 (2008): 47–58; Melinda D. Chambers et al., "Genetic Structure of Black Abalone (*Haliotis cracherodii*) Populations in the California Islands and Central California Coast: Impacts of Larval Dispersal and Decimation from Withering Syndrome," *Journal of Experimental Marine Biology and Ecology* 331 (2006): 173–185; Rob Burton, interview, April 27, 2010.
39 Hunter S. Lenihan, "Assessing Withering Syndrome Resistance in California Black Abalone: Implications for Conservation and Restoration," Research Completion Report, UC San Diego: California Sea Grant College Program (January 29, 2009), 2–3;

researchers at NMFS continue to research propagation techniques for black abalone, NMFS, Protected Resources Division, "Black Abalone (*Haliotis cracherodii*), Five-Year Status Review: Summary and Evaluation," 2018, 22.

40 Center for Biological Diversity, "Petition to List the Black Abalone (*Haliotis cracherodii*) as Threatened or Endangered under the Endangered Species Act," December 21, 2006, https://www.biologicaldiversity.org/species/invertebrates/black_abalone/pdfs/Black-Ab-Petition-12-21-06.pdf.

41 Butler et al., "Status Review," ix; "Endangered and Threatened Wildlife and Plants; Endangered Status for Black Abalone," *Federal Register* 74, no. 9 (2009): 1937–1946.

42 V. R. Restrepo and J. E. Powers, "Precautionary Control Rules in U.S. Fisheries Management: Specification and Performance," *ICES Journal of Marine Science* 56 (1999): 846–852.

43 M. L. Weber and B. Heneman, *Guide to California's Marine Life Management Act* (Common Knowledge Press, 2000), ii–xxiv; CDFG, Marine Protected Areas website, https://www.wildlife.ca.gov/Conservation/Marine/MPAs.

44 Rogers-Bennett, interview, 2011.

45 Haaker, interview, 2010.

46 Konstantin A. Karpov et al., "Serial Depletion and the Collapse of the California Abalone (*Haliotis* spp.) Fishery," in *Workshop on Rebuilding Abalone Stocks in British Columbia*, ed. Alan Campbell, Canadian Special Publication in Fisheries and Aquatic Science 130 (NRC Research Press, 2000), 11–24.

47 Rogers-Bennett, interview, 2011.

48 Russ Babcock and John Keesing, "Fertilization Biology of the Abalone *Haliotis laevigata*: Laboratory and Field Studies," *Canadian Journal of Fisheries and Aquatic Science* 56 (1999): 1677; Rogers-Bennett, interview, 2011.

49 CDFG, *ARMP*, iii–iv; 6-4–6-5.

50 Restrepo and Powers, "Precautionary Control," 846–852; Rogers-Bennett, interview, 2011; Ian Taniguchi, interview, April 30, 2010.

51 Laura Rogers-Bennett, interview, August 14, 2019.

52 CDFG, *ARMP*, 7-4–7-17.

53 Mia J. Tegner et al., "Climate Variability, Kelp Forests, and the Southern California Red Abalone Fishery," *Journal of Shellfish Research* 20, no. 2 (2001): 760.

54 CDFG, *ARMP*, 6-1–6-32; Rogers-Bennett, interview, 2011.

55 CDFG, *ARMP*, Appendix G, 1–51.

56 CFGC, video meeting minutes, November 3, 2005; CDFG, *ARMP*, Appendix H, 8–9.

57 Jeremy D. Prince, "Combating the Tyranny of Scale for Haliotids: Micro-Management for Microstocks," *Bulletin of Marine Science* 76, no. 2 (2005): 557–577; Jeremy D. Prince et al., "The Novel Use of Harvest Policies and Rapid Visual Assessment to Manage Spatially Complex Abalone Resources (Genus *Haliotis*)," *Fisheries Research* 94 (2008): 330–338.

58 CFGC, video meeting minutes, November 3, 2005.

59 CFGC, video meeting minutes, November 3, 2005.

60 CFGC, video meeting minutes, December 8, 2005.

61 CFGC, video meeting minutes, December 8, 2005.

62 CFGC, video meeting minutes, December 8, 2005; in the final *ARMP*, the SMI fishery option was called "Alternative 1."

63 Kenneth R. Weiss, "Abalone Fishery off Southland May Reopen," *Los Angeles Times*, December 23, 2005; CFGC, video meeting minutes, December 8, 2005.

64 Rocky Daniels, "Marine Resource Issues Commercial Abalone Fishery Reopening?" Rocky Pages blog, http://www.sonic.net/~rocky/icommercial.htm, accessed March, 23, 2010; Weiss, "Abalone Fishery"; Steve Benavides, "Abalone Update," *Underwater Flash* (newsletter of the Los Angeles County Underwater Photography Society), November 2005, 5–6.

65 Weiss, "Abalone Fishery."

66 Scott McCreary et al., "The Abalone Advisory Group Report on Management Options for Establishing a Potential Red Abalone Fishery at San Miguel Island," prepared by CONCUR, January 29, 2010, Appendix C, Meeting Minutes and Key Outcome Memos

(September 2006–September 2009); Taniguchi, interview, 2010. Documents related to the San Miguel Island Red Abalone Fishery Consideration Process can be found at https://www.wildlife.ca.gov/Conservation/Marine/ARMP/San-Miguel-Island.

67 Alistair Brand, "Diving for Delicacies and Dollars," *Santa Barbara Independent*, June 26, 2008; "The Abalone Advisory Group Report," Appendix A, Fishery Management Options; CDFG, "San Miguel Island Red Abalone Fishery Considerations: Report to the Marine Resources Committee" (December 2012), 5. http://www.fgc.ca.gov/meetings/committees/MR%20Recommendation%20for%20SMI%20Red%20Ab%20fishery.pdf.

68 CFGC, video meeting minutes, March 3, 2010.

69 Matt Kettmann, "Alleged Abalone Poachers Caught," *Santa Barbara Independent*, December 13, 2012; "Poacher Convicted," *Santa Barbara Independent*, April 24, 2014.

70 CFGC, video meeting minutes, February 13, 2013; CDFG, "San Miguel Fishery Considerations," 3; $1 million reported by Deputy Director Sonke Mastrup at CFGC meeting, March 3, 2010.

71 CFGC, video meeting minutes, February 13, 2013.

72 Rogers-Bennett, interview, 2011.

73 Tegner et al., "Climate Variability," 761.

74 CDFG, *ARMP*, "Historical Summary," A14; Karpov et al., "Serial Depletion," 19.

75 Alistair J. Hobday and Mia J. Tegner, "The Warm and the Cold: Influence of Temperature and Fishing on Local Population Dynamics of Red Abalone," *CalCOFI Report* 43 (2002): 93–94.

76 L. Ignacio Vilchis et al., "Ocean Warming Effects on Growth, Reproduction and Survivorship of Southern California Abalone," *Ecological Applications* 15, no. 2 (2005): 469–480; James D. Moore et al., "Green Abalone *Haliotis fulgens* Infected with the Agent of Withering Syndrome Do Not Express Disease Signs under a Temperature Regime Permissive for Red Abalone, *Haliotis rufescens*," *Marine Biology* 156, no. 11 (2009): 2325–2330.

77 Laura Rogers-Bennett et al., "Response of Red Abalone Reproduction to Warm Water, Starvation, and Disease Stressors: Implications of Ocean Warming," *Journal of Shellfish Research* 29, no. 3 (2010): 599–611.

78 Melissa Neuman et al., "Overall Status and Threats Assessment of Black Abalone (*Haliotis cracherodii* Leach, 1814) Populations in California," *Journal of Shellfish Research* 29 (2010): 577–586.

79 Joan Kleypas and Chris Langdon, "Overview of CO_2-Induced Changes in Seawater Chemistry," in *Proceedings of the Ninth International Coral Reef Symposium*, Bali, Indonesia, v. 2 (2000), 1085–1089.

80 Richard A. Feely et al., "Impact of Anthropogenic CO_2 on the $CaCO_3$ System in the Oceans," *Science* 305 (2004): 362–366; James C. Orr et al., "Anthropogenic Ocean Acidification over the Twenty-First Century and Its Impact on Calcifying Organisms," *Nature* 437 (2005): 681–686.

81 Victoria J. Fabry et al., "Impacts of Ocean Acidification on Marine Fauna and Ecosystem Processes," *ICES Journal of Marine Science* 65, no. 3 (2008): 414–432; N. Bednaršek et al., "Extensive Dissolution of Live Pteropods in the Southern Ocean," *Nature Geosciences* 5 (2012): 881–885.

82 Alan Barton et al., "The Pacific Oyster, *Crassostrea gigas*, Shows Negative Correlation to Naturally Elevated Carbon Dioxide Levels: Implications for Near-Term Ocean Acidification Effects," *Limnology and Oceanography* 57, no. 3 (2012): 698–710; Elizabeth Grossman, "Northwest Oyster Die-Offs Show Ocean Acidification Has Arrived," *Yale Environment 360*, November 21, 2011, https://e360.yale.edu.

83 Maria Byrne et al., "Unshelled Abalone and Corrupted Urchins: Development of Marine Calcifiers in a Changing Ocean," *Proceedings of the Royal Society of Biological Sciences* 278 (2011): 2376–2883.

84 Ryan N. Crim et al., "Elevated Seawater CO_2 Concentrations Impair Larval Development and Survival in Endangered Northern Abalone (*Haliotis kamtschatkana*)," *Journal of Experimental Biology and Ecology* 400 (2011): 272–277.

85 Ryo Kimura et al., "Effects of Elevated pCO_2 on the Early Development of the Commercially Important Gastropod, Ezo Abalone, *Haliotis discus hannai*," *Fisheries Oceanography* 20, no. 5 (2011): 357–366.
86 Orr et al., "Anthropogenic Ocean," 685; Ilsa B. Kuffner et al., "Decreased Abundance of Crustose Coralline Algae Due to Ocean Acidification," *Nature Geoscience* 1 (2008): 114–117.

CHAPTER 14

1 "White Abalone Recovery Plan (*Haliotis sorenseni*)," NMFS, Office of Protected Resources, Long Beach, CA (October 2008), 1-28; NOAA, "Endangered Species: File No. 14344," *Federal Register* 76, no. 66 (2011): 19052–19053; Kristin Aquilino, interview, December 8, 2014.
2 Aquilino, interview, 2014; Laura Rogers-Bennett et al., "Implementing a Restoration Program for the Endangered White Abalone (*Haliotis sorenseni*) in California," *Journal of Shellfish Research* 35, no. 3 (2016): 614; "Saving the Endangered White Abalone: Speeding Up a Snail's Pace," lecture by Kristin Aquilino, July 22, 2014, CDFW, Conservation Lecture Series; David Witting, correspondence, October 8, 2019.
3 Rogers-Bennett et al., "Implementing," 2016, 612–613; Aquilino, interview, 2014; Jim Moore, interview, March 17, 2011.
4 Aquilino, interview, 2014.
5 Kevin L. Stierhoff et al., "On the Road to Extinction? Population Declines of the Endangered White Abalone, *Haliotis sorenseni*," *Biological Conservation* 152 (2012): 46–52; Rogers-Bennett et al., "Implementing," 611; Cynthia A. Catton et al., "Population Status Assessment and Restoration Modeling of White Abalone *Haliotis sorenseni* in California," *Journal of Shellfish Research* 35, no. 3 (2016): 593–599.
6 Dave Witting, interview, September 26, 2019; "Diving Deeper into Abalone, Dedicated Scientists and Volunteers on a Mission to Better Understand Endangered Species," NOAA Fisheries, West Coast Region, Recent Stories, Spring 2015, https://archive.fisheries.noaa.gov/wcr/stories/2015/05112015_abalone_citizen_science.html.
7 Melissa Neuman, interview, June 19, 2019; Jim Moore, interview, June 25, 2019; Deborah Sullivan Brennan, "Lonely Abalone Looking for Love: Can Captive Breeding Save Endangered Shellfish?" *San Diego Union Tribune*, June 16, 2017; "Department Name Change Effective Tomorrow," *CDFW News*, December 31, 2012. The department's name changed in 2013 from "Fish and Game" to "Fish and Wildlife," reflecting an important change in the legislature's direction for management of California's animals. The Fish and Game Commission's name did not change.
8 Lindsay Hoshaw, "The Extraordinary Effort to Save the White Abalone," KQED Science, March 17, 2017, https://www.kqed.org/science/1339079/delicious-and-nearly-extinct-can-white-abalone-be-saved.
9 Kristin Aquilino, interview, June 18, 2019.
10 Rogers-Bennett et al., "Implementing," 614–617; Neuman, interview, 2019; Chris Yates (Assistant Regional Administrator for Protected Resources, NOAA), interview, October 15, 2019.
11 Kat Kerlin, "Endangered White Abalone Program Yields Biggest Spawning Success Yet Millions of Eggs Bring Program 1 Step Closer to Saving Species," in *UC Davis Environment News*, April 25, 2019, https://www.ucdavis.edu/news/endangered-white-abalone-program-yields-biggest-spawning-success-yet/.
12 Aquilino, interview, 2019.
13 Rogers-Bennett et al., "Implementing," 612–613; Ian Taniguchi, interview, February 19, 2015; Heather Burdick (The Bay Foundation), interview, July 1, 2019.
14 NMFS, West Coast Region, "White Abalone (*Haliotis sorenseni*), Five-year Status Review: Summary and Evaluation" (July 2018), 4–5; Jim Moore, interviews, February 7, 2014, and June 25, 2019; Aquilino, interview, 2019; Neuman, interview, 2019; Witting, interview, 2019.
15 Moore, interview, 2019.

16 Henry S. Carson et al., "The Survival of Hatchery-Origin Pinto Abalone *Haliotis kamts-chatkana* Released into Washington Waters," *Aquatic Conservation: Marine and Freshwater Ecosystems* (2019): 1–18.

17 Neuman, interview, 2019.

18 Kat Kerlin, "Saving Endangered Species in a Changing Ocean," *Washington Post*, September 25, 2017; Aquilino, interview, 2019.

19 Hoshaw, "Extraordinary Effort."

20 Witting, interview, 2019.

21 Lynn Lilliston, "Conservationists Fighting Drive against Sea Otters," *Los Angeles Times*, March 30, 1970; Biliana Cicin-Sain et al., "Social Science Perspectives on Managing Conflicts between Marine Mammals and Fisheries," Marine Policy Program, Marine Science Institute, UC Santa Barbara and UC Cooperative Extension, 1982; US Fish and Wildlife Service, "Final Revised Recovery Plan for the Southern Sea Otter (*Enhydra lutris nereis*)," Portland, OR, 2003, 12–13, 19.

22 John Butler et al., "Status Review Report for Black Abalone (*Haliotis cracherodii* Leach, 1814)," NMFS (2009), 13; Glenn VanBlaricom, interview, October 24, 2015.

23 Lisa M. Crosson et al., "Abalone Withering Syndrome: Distribution, Impacts, Current Diagnostic Methods and New Findings," *Diseases of Aquatic Organisms* 108 (2014): 261–270; Lisa M. Crosson et al., "Influence of Rickettsial Pathogens on Endangered Black Abalone: Differential Susceptibility and Host Response," PowerPoint presentation, nd; Lisa M. Crosson et al., "A Transcriptomic Approach in Search of Disease Resistance in Endangered Black Abalone," PowerPoint, September 2012 (both PowerPoints in author collection); Lisa Crosson, interview, November 20, 2015; Drew Harvell, *Ocean Outbreak: Confronting the Rising Tide of Marine Disease* (University of California Press, 2019), 74–75.

24 Crosson et al., "Abalone Withering Syndrome," 267; Harvell, *Ocean Outbreak*, 75; Crosson, interview, 2015.

25 Carolyn S. Friedman and Lisa M. Crosson, "Putative Phage Hyperparasite in the Rickettsial Pathogen of Abalone, *Candidatus Xenohaliotis californiensis*," *Microbial Ecology* 64, no. 4 (2012): 1064–1072; Crosson et al., "Abalone Withering Syndrome," 261–270; Carolyn S. Friedman et al., "Reduced Disease in Black Abalone Following Mass Mortality: Phage Therapy and Natural Selection," *Frontiers in Microbiology* 5 (2014): 78.

26 Harvell, *Ocean Outbreak*, 74; Moore, interview, 2019.

27 Kevin D. Lafferty and Leah R. Gerber, "Good Medicine for Conservation Biology: The Intersection of Epidemiology and Conservation Theory," *Conservation Biology* 16, no. 3 (2002): 593–604.

28 Brian B. Hatfield, "The Translocation of Sea Otters to San Nicolas Island: An Update," *Sixth California Islands Symposium* (Institute for Wildlife Studies, Arcata, California, 2005), 473–474.

29 Brian B. Hatfield et al., "California Sea Otter (*Enhydra lutris nereis*) Census Results, Spring 2018," USGS Data Series 1097 (2018), 4–5.

30 Peter Raimondi et al.," Evaluating Potential Conservation Conflicts between Two Listed Species: Sea Otters and Black Abalone," *Ecology* 96, no. 11 (2015): 3102–3107.

31 NMFS, West Coast Region, Protected Resources Division, "Black Abalone, Five-Year Status Review" (July 2018), 11; Glenn VanBlaricom, interview, October 26, 2015; Neuman, interview, 2014.

32 NMFS, "Black Abalone, Five-Year Status Review," 5, 8.

33 CINP, "Black Abalone Regain Lost Ground," March 15, 2017, https://www.nps.gov/articles/black-abalone-regain-lost-ground.htm.

34 L. Ignacio Vilchis et al., "Ocean Warming Effects on Growth, Reproduction, and Survivorship of Southern California Abalone," *Ecological Applications* 15, no. 2 (2005): 477.

35 Nancy Caruso, interview, April 30, 2011; Pat Brennan, "Abalone Part of Lessons in Classroom before Release," *Orange County Register*, December 5, 2012; Jennifer Erickson, "Growing Abalone and Cultivating Ocean Stewards," *Laguna Beach Independent*, March 23, 2011; Pat Brennan, "O.C. Classrooms Saving Troubled Species?" *Orange County Register*, February 9, 2011.

36 Nancy Caruso, interview and field trip, April 30, 2011.

37 Nancy Caruso, interview, December 5, 2014.

38 Fred Wendell, "California Department of Fish and Game's Response to the Sabellid Worm Infestation," in *Biological Invasions in Aquatic Ecosystems: Impacts on Restoration and Potential for Control, Proceedings of a Workshop* (April 25, 1998, Sacramento), ed. A. N. Cohen and S. K. Webb (San Francisco Estuary Institute, 2002), 8.

39 Nancy Caruso, interview, October 26, 2010.

40 Taniguchi, interview, 2015.

41 Taniguchi, interview, 2015.

42 K. M. Gruenthal, "Development and Application of Genomic Tools to the Restoration of Green Abalone in Southern California," *Conservation Genetics* 15, no. 1 (2014): 109–121; Melinda D. Chambers et al., "Genetic Structure of Black Abalone (*Haliotis cracherodii*) Populations in the California Islands and Central California Coast: Impacts of Larval Dispersal and Decimation from Withering Syndrome," *Journal of Experimental Marine Biology and Ecology* 331 (2006): 173–185.

43 Pat Brennan, "Earth Day: Kelp, Abalone, Oceans of Science," *Orange County Register*, April 21, 2012.

44 Caruso, interview, December 5, 2014; Aaron Orlowski, "We Ate Our Most Popular Mollusk to the Point of Near-Extinction," *Oregon County Register*, December 1, 2014.

45 Caruso, interviews, December 5, 2014, and July 1, 2019.

46 Laylan Connely, "Nostalgic for California Abalone? Why They Disappeared, and How 1 Project Aims to Bring Them Back," *Orange County Register*, July 19, 2018.

47 Witting, interview, 2019.

48 Konstantin Karpov, "Depth Refuge: Management That Works," *Kelp Forest Newsletter* (May 1998): 1, CDFG, Marine Region.

49 CDFG, *ARMP*, "Historical Summary of Laws and Regulations," Appendix A, A5–A6; CDFG, *ARMP*, G18; Bruce Watkins, "California's Changing Abalone Regulations: The Science behind the Law," *California Diving News* (January 2002).

50 CFGC, Revised Initial Statement of Reasons for Regulatory Action, Amend Sec. 29.15, Title 14, Re: Abalone (August 5, 2011), 2.

51 Pierre De Wit et al., "Forensic Genomics as a Novel Tool for Identifying the Causes of Mass Mortality Events," *Nature Communications* 5, no. 3652 (2014): 1–4; "SCAN Opposes Reduction in Annual Abalone Bag Limit—F&G Commission Agrees," *SCAN Newsletter*, October 25, 2014.

52 CFGC, video meeting minutes, February 10, 2016; Laura Rogers-Bennett and Cynthia A. Catton, "Marine Heat Wave and Multiple Stressors Tip Bull Kelp Forest to Sea Urchin Barrens," *Scientific Reports* 9 (2019): 1–9.

53 Cynthia Catton, et al., "'Perfect Storm' Decimates Northern California Kelp Forests," *CDFW Marine Management News*, March 30, 2016; Laura Rogers-Bennett, "Perfect Storm: Multiple Stressors Push Kelp Forest beyond Tipping Point," lecture, July 28, 2019, CDFW, Conservation Lecture Series.

54 Catton, et al., "'Perfect Storm.'"

55 CFGC, Final Statement of Reasons for Regulatory Action, Amend Sec. 29.15, Title 14, Re: Abalone (October 1, 2013), 3–39.

56 CFGC, Final Statement, 2013.

57 CFGC, Final Statement, 2013; Joshua Russo (Watermen's Alliance), interview, August 5, 2019.

58 CFGC, Final Statement, 2013.

59 CDFW, Press Release, "Recreational Red Abalone Season to Open along Northern California Coast," March 20, 2015; CDFW, Marine Region, "2014 By the Numbers," 2; CDFW, Marine Region, "2015 By the Numbers," 3.

60 CDFW, Recreational Abalone Advisory Committee, meeting minutes, November 14, 2015, https://www.wildlife.ca.gov/Conservation/Marine/Invertebrates/RAAC; Michael L. Weber et al., *Guide to California's Marine Life Management Act*, 2nd ed. (California Wildlife Foundation, 2018), 15.

61 CFGC, video meeting minutes, February 10, 2016.

62 Mary Callahan, "North Coast Kelp Beds 'Like a Desert' This Year," *Santa Rosa Press Democrat,* July 21, 2016; Ross Clark, "Earth Matters: The Loss of Our Kelp Forests," *Santa Cruz Sentinel,* February 1, 2018.

63 Callahan, "North Coast Kelp."

64 CFGC, meeting packet, Abalone Emergency, Staff Summary for August 16, 2017, Item No. 18.

65 Cynthia Catton, "Recent Kelp Loss Impacts Ecosystem Health and Abalone Fishery in Northern California," presentation, CFGC, Marine Resources Committee, November 15, 2016; CFGC, video meeting minutes, December 7, 2016.

66 Tara Duggan, "Abalone Divers, State Agency at Odds over Proposed Restrictions," *San Francisco Chronicle,* December 1, 2016; CFGC, meeting packet, Abalone Emergency, Staff Summary for December 7–8, 2016, Item No. 10; Tara Duggan, "California Sets Stricter Limits on Abalone," *San Francisco Chronicle,* December 7, 2016.

67 CFGC, Marine Resources Committee meeting, Committee Staff Summary, July 20, 2017; Sonke Mastrup, "Red Abalone Fishery Update," presentation, in CFGC, Marine Resources Committee meeting, Committee Staff Summary, July 20, 2017.

68 CFGC, Initial Statement of Reasons for Regulatory Action, Amend Sec. 29.15, Title 14, Re: Abalone (September 12, 2017), 1–23; Clark, "Earth Matters"; Mark H. Carr to CFGC, letter, regarding status of red abalone stock in northern California, November 22, 2017, in CFGC, meeting packet, Recreational Abalone, Staff Summary for December 6–7, 2017, Item No. 22.

69 Mastrup, "Red Abalone Fishery Update"; Mary Callahan, "More Limits Foreseen for California Abalone Fishery as Scientist Raises Alarm," *Santa Rosa Press Democrat,* June 22, 2017; CFGC, meeting packet, Abalone Emergency, Staff Summary for August 16, 2017, Item No. 18.

70 Duggan, "Abalone Divers"; CFGC, video meeting minutes, December 7, 2016; Jack Shaw (Abalone Working Group) to CFGC, November 22, 2016, letter, regarding Abalone Emergency Regulations and Fishery Management Plan, in CFGC, meeting packet, Abalone Emergency, Staff Summary for December 7–8, 2016, Item No. 10.

71 Tom Dempsey (The Nature Conservancy) to CFGC, letter, regarding Agenda Item 6B, Red Abalone—Update on Fishery Management Plan Development, July 7, 2017, in CFGC, meeting packet, Red Abalone, Staff Summary for July 20, 2017, Item No. 6.

72 CFGC, meeting packet, Recreational Abalone, Staff Summary for August 16, 2017, Item No. 19; CFGC meeting packet, Recreational Abalone, Staff Summary for December 6–7, 2017, Item No. 22.

73 CFGC, meeting, December 7, 2017, video minutes; Joshua Russo, interview, 2019.

74 CFGC, meeting, December 7, 2017, video minutes; Tara Duggan, "Abalone Diving Banned Next Year to Protect Population on Brink of Collapse," *San Francisco Chronicle,* December 7, 2017.

75 CFGC, video meeting minutes, December 7, 2017.

76 CFGC, video meeting minutes, December 7, 2017.

77 CFGC, video meeting minutes, December 7, 2017; CFGC, Final Statement of Reasons for Regulatory Action, Amend Sec. 29.15, Title 14, Re: Abalone Regulations (January 25, 2018), 1–7; CFGC, Final Statement of Reasons for Regulatory Action, Amend Sec. 29.15, Title 14, Re: Recreational Take of Red Abalone (January 2, 2019), 1–6; Mary Callahan, "California's Abalone Fishery to Be Closed Next Year to Protect Crashing Fishery," *Santa Rosa Press Democrat,* December 7, 2017; Michelle Blackwell, "Abalone Season Halted until 2021 as Storm Swells Toss Mollusks on Shore," *Fort Bragg Advocate-News,* January 10, 2019.

78 Rietta Hohman et al., "Sonoma-Mendocino Bull Kelp Recovery Plan," Greater Farallones National Marine Sanctuary and CDFW (2019), ix–xv, https://farallones.org/wp-content/uploads/2019/06/Bull-Kelp-Recovery-Plan-2019.pdf.

79 Hohman et al., "Sonoma-Mendocino," 27–28; Joshua Russo, interview.

80 Kimberly Wear, "A Miracle Didn't Happen: Abalone Season Could Be Shuttered for Another Two Years," *North Coast Journal,* December 6, 2018.

81 Rogers-Bennett and Catton, "Marine Heat Wave," 1–9; Laura Rogers-Bennett, interview, August 14, 2019.

CONCLUSION

1 Curtis G. Berkey and Scott W. Williams, "California Indian Tribes and the Marine Life Protection Act: The Seeds of a Partnership to Preserve Natural Resources," *American Indian Law Review* 43 no. 2 (2019), 338–350.

Index